Lasers in Medicine and Surgery
AN INTRODUCTORY GUIDE

Third Edition

G. T. Absten
BSc, MA

Clinical Instructor, The Ohio State University College of Medicine and President of Advanced Laser Services Corporation, Columbus, Ohio

and

S. N. Joffe
BSc, MBChB, MD, FRCS, FCS, FACS

Esteemed Quondam Professor of Surgery at the University of Cincinnati Medical Center and President of Laser Centers of America, Inc.

CHAPMAN & HALL MEDICAL
London · Glasgow · New York · Tokyo · Melbourne · Madras

Published by Chapman & Hall, 2–6 Boundary Row, London SE1 8HN

Chapman & Hall, 2–6 Boundary Row, London SE1 8HN, UK

Blackie Academic & Professional, Wester Cleddens Road, Bishopbriggs, Glasgow G64 2NZ, UK

Chapman & Hall Japan, Thomson Publishing Japan, Hirakawacho Nemoto Building, 6F, 1-7-11 Hirakawa-cho, Chiyoda-ku, Tokyo 102, Japan

Chapman & Hall Australia, Thomas Nelson Australia, 102 Dodds Street, South Melbourne, Victoria 3205, Australia

Chapman & Hall India, R. Seshadri, 32 Second Main Road, CIT East, Madras 600 035, India

Distributed exclusively in the USA by Laser Centers of America Inc., 1111 St Gregory St, Cincinnati, Ohio 45202, USA

First edition 1985
Second edition 1989
Third edition 1993

© 1993 G.T. Absten and S.N. Joffe

Phototypeset in 10/12 Times by Intype, London
Printed in Great Britain by The University Press, Cambridge

ISBN 0 412 46890 5

British Library Cataloguing in Publication Data

Absten, Gregory, T.
    Lasers in medicine and surgery – 3rd ed.
    1. medicine. Use of lasers
    I. Title   II. Joffe, Stephen N.
    610'.28

    ISBN 0 412 35560 4   0 412 30870 3 (Pb)

Library of Congress Cataloging-in-Publication Data

Absten, Gregory T.
    Lasers in medicine: an introductory guide/
      Gregory T. Absten,
    Stephen N. Joffe. – [3rd ed.]
        p.   cm.
    Bibliography: p.
    Includes index.
    ISBN 0 412 35560 4   0 412 30870 3 (Pb)
    1. Lasers in medicine and surgery.   I. Joffe, Stephen N.   II. Title.
R857.L37A27   1988
610'.28–dc19   88-10843
                                                      CIP

Contents

Acknowledgements

We would like to thank all the physicians, nurses, and others we have worked with over the years, in hospitals throughout the United States, Europe and the Far East, for their professional support and encouragement. No matter how much we teach, we always feel as if we are learning new things from those being instructed.

GTA: I especially want to acknowledge my wife Melanie and six children, Heather, Eric, Kimberly, Nikki, Lindsay and, most recently, baby Emily for accepting me as the 'laser-head' I am.

SNJ: I want to thank my wife Sandra and two children, Heidi and Craig, for assisting me in efforts to increase laser utilization in the medical and surgical fields to provide better quality of patient care.

G.T. Absten
S.N. Joffe

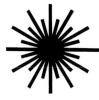

Preface

In every area of human endeavor, technology has opened the door for new advancements to occur. Much of the progress in surgery and medicine over the last few years is due, in large part, to new technological instruments made available to clinicians and researchers. Laser surgery is an ever expanding technological discipline in surgery and medicine that is contributing to a broad and rapid expansion of both diagnostic and treatment procedures.

Laser is to light what music is to noise. Those physicians who wish to be most successful in the application of this technology, to the benefit of their patient, will learn of the subtle interactions of light with tissue.

No technology is good or bad in itself. It is only in the choices we make, in when and how to apply that technology, that it gains its moral value. The use of lasers has some very definite advantages in the surgical and medical treatment of a variety of disorders.

At the same time we must all be careful to not perpetrate the myth of lasers in surgery. Vastly overstated claims of the value of 'laser surgery' have been held out to the general public, resulting in health care being sought on the basis of laser availability.

It is important to remember that, in the practice of medicine, it is the physicians who are the prime determinants of the quality of health care, and not the tools they choose to use. The laser can be a wonderful tool when used by an experienced, well trained physician.

This book has been written to provide a basic introduction to the use of lasers in medicine and surgery – what they are, how they work, and what they can do for the patient. It assumes only a basic scientific background for the reader, and has many simple and clear diagrams. It should be excellent as an overview of the activity in lasers in medicine to physicians, medical students, nurses, biomedical engineers, patients and the interested public. The science of light is a fascinating topic which is easily understood from a conceptual point of view.

An encyclopedic glossary of laser terms is included which, in itself, provides a quick, easy introduction to medical laser concepts.

The physical principle on which lasers are based developed from Einstein's theories in the early 1900s, though the first laser device was not produced until 1960 by Maiman. Einstein had a quest – to show that the four basic forces of the universe are simply different faces of the same singular force. The single force of the universe – that is the ultimate quest of all of physics, and of philosophy.

Glossary

Ablation	volume removal of tissue by vaporization.
Absorption	uptake of light energy by tissue, converting it into heat.
Absorption coefficient	factor describing light's ability to be absorbed. Optical properties of different tissues alter the absorption.
Active medium	(laser medium) the material used to emit the laser light.
Aiming beam	a HeNe laser (or other light source) used as a guide light. Used coaxially with infrared or other invisible light.
Amplitude	the maximum height of a wave. Implies power.
Argon	the gas used as a laser medium. It emits blue/green light at 488 and 515 nm.
Articulated arm	a CO_2 laser delivery device consisting of hollow metal tubes with joints which allow the 'arm' to move. Mirrors are located at each joint to reflect the laser beam.
Attenuation	decreasing the intensity (power) of light as it passes through a medium.
Biostimulation	the use of low power light (milliwatts), usually laser, to stimulate metabolic activity on a subcellar level. Experimentally examined for pain relief and wound healing.
Carbon dioxide	(CO_2) molecule used as a laser medium. Emits far infrared light at 10 600 nm

(10.6 μm). Lasers are made as sealed tube, or flowing gas units.

Cautery — achieving hemostasis of bleeding vessels, usually by heat from laser or electrosurgical units. Contrasts with laser induced protein coagulation.

Chromophore — optically active (colored) material in tissue which acts as the target for laser light.

Coagulation — destruction of tissue by heat without physically removing it.

Coherence — orderliness of wave patterns by being in phase in time and space.

Combiner mirror — the mirror in a laser which combines two or more wavelengths into a coaxial beam, i.e. CO_2 and HeNe beams.

Contact probe — synthetic ceramic material, like sapphire, used with laser fibers to allow touch of tissue with the probe, intensifying its effects, and allowing cutting, vaporizing, or coagulation of tissue at relatively low powers and high degree of control.

Continuous wave — (CW) constant, steady state delivery of laser power.

Collimation — ability of the laser beam to not spread (low divergence) with distance.

Dichroic filter — filter that allows selective transmission of colors.

DHE — dihematoporphyrin ether. A photosensitizing agent used in PDT. DHE is a more refined form of HpD.

Diffuser — an optical device or material that homogenizes the output of light causing a very smooth, even distribution over the area affected.

Dosimetry — measuring the amount (joules) and intensity (watts/cm^2) of light delivered to tissue.

Electron	negatively charged particle of an atom.
Electromagnetic spectrum	the span of frequencies (wavelengths) considered to be light – from radio and TV waves to gamma and cosmic rays.
Endoscope	an instrument inserted into the body through an orifice (either existing or surgical) that allows viewing and manipulation of tissue. May be rigid or flexible.
Energy	expressed in joules (watt-seconds).
Excimer	'excited dimer'. A gas mixture used as the basis of lasers emitting ultraviolet light.
Excitation	energizing a material into a state of population inversion.
Femtoseconds	10^{-15} seconds. Shorter than picosecond or nanosecond.
Fiberoptics	a system of flexible quartz or glass fibers with internal reflective surfaces that pass light through thousands of glancing reflections. Many hundreds or thousands of individual fibers are needed to transmit an image, but only single fibers are used to transmit laser light during treatment.
Focal point	that distance from the focusing lens where the laser beam has the smallest spot diameter and hence greatest intensity.
Gated pulse	a discontinuous burst of laser light, made by timing (gating) a continuous wave output – usually in fractions of a second.
Gaussian curve	normal statistical curve showing a peak with even distribution on either side. May either be a sharp peak with steep sides, or a blunt peak with shallower sides. Used to show power distribution in a beam. The concept is important in controlling the geometry of the laser impact.
Hemostasis	the ability to stop bleeding.

HeNe	helium neon. A laser producing low power (milliwatts) red light (630 nm) used as a guide light for infrared lasers, or experimentally for biostimulation.
Hologram	a three dimensional picture made by interference patterns created by the coherence of laser light. Created as transmission, reflection or integral holograms.
HpD	hematoporphyrin derivative. A photosensitizing drug used with photodynamic therapy as a treatment for cancer.
Impact size	the size crater or width of incision left by a laser impact. Related to spot size of the beam, except impact size varies depending on how the energy is applied.
Ionizing radiation	radiation commonly associated with X-Ray, that is of a high enough energy to cause DNA damage with no direct, immediate thermal effect. Contrasts with non-ionizing radiation of surgical lasers.
Irradiance	*see* Power density.
Joule	a unit of energy. Laser powers are sometimes described in joules per second. A power of 1 joule per second is known as 1 watt and is the rate of energy delivery.
KTP	potassium titanyl phosphate. A crystal used to change the wavelength of a Nd:Yag laser from 1060 nm (infrared) to 532 nm (green).
Laser	Light Amplification by the Stimulated Emission of Radiation. A device that produces intense beams of pure colors of light.
Laser medium	(active medium) material used to emit the laser light and for which the laser is named.
Laser surgeon	no such thing; but surgeons do use lasers to advantage as surgical instruments.
Metastable state	the state of an atom, just below a higher

excited state, which an electron occupies momentarily before destabilizing and emitting light.

Micromanipulator (microslad)

device attached to a microscope that controls delivery of the laser beam into the microscopic field of view. In non-ophthalmic surgery, they are most commonly used with CO_2 lasers, then with argon and KTP, and least with Nd:Yag lasers.

Microprocessor

a digital chip (computer) that operates and monitors some lasers.

Mode

a term used to describe how the power of a laser beam is distributed within the geometry of the beam. Also used to describe the operating mode of a laser such as continuous or pulsed.

Mode-locking

a process similar to Q-switching except that the pulses produced are even shorter (about 10^{-12} second) and emerge in short trains of pulses instead of singularly. It is usually achieved with a dye cell.

Monochromaticity

waves are monochromatic when they are all of the same wavelength (color).

Nanometer

abbreviated nm – a measure of length. One nm equals 10^{-9} meter, and is the usual measure of light wavelength. Visible light ranges from about 400 nm in the purple to about 750 nm in the deep red.

Nanosecond

10^{-9} (one billionth) of a second. Longer than a picosecond or femtosecond, but shorter than a microsecond. Associated with Q-switched ophthalmic Nd:Yag lasers.

Neodymium

the rare earth element that is the active element in a Nd:Yag laser.

Nd:Yag

neodymium:yttrium aluminum garnet. A mineral crystal used as a laser medium to produce 1060 nm light.

Nonlinear effect

not a normal, linear temperature rise induced by laser. Refers to the plasma 'spark' and snap created by the Q-switched Nd:Yag laser.

Optical breakdown

plasma formation by stripping electrons off atoms/molecules. Caused by high laser energy densities and used to create a 'spark'. Used in ophthalmology with Q-switched or mode-locked Nd:Yag lasers to cut membranes.

Optical cavity (resonator)

space in between the laser mirrors where lasing action occurs.

Output coupler

the partially transmissive mirror that allows laser output from the optical cavity.

PDT

photodynamic therapy. The use of photosensitizing drugs, activated by certain pure colors of light produced by the laser, to achieve selective tissue destruction. Its current major use is investigationally as a selective treatment for cancer.

Phase

waves are in phase with each other when all the troughs and peaks coincide and are 'locked' together. The result is a reinforced wave of increased amplitude (brightness).

Photocoagulation

tissue coagulation caused by light (laser).

Photodisruption

creating an acoustical shock wave, through Q-switching or mode-locking, to gently 'snap' apart membranes. This is a 'cold cutting' technique with laser. Ophthalmologists use the Q-switched Nd:Yag to photodisrupt an opacified posterior capsule secondary to cataract surgery.

Photon

the basic particle of light.

Picosecond

10^{-12} seconds. Longer than a femtosecond but shorter than a nanosecond. Associated with mode-locked ophthalmic Nd:Yag lasers.

Plasma

the fourth state of matter in which electrons

have been stripped off the atoms. The extremely high internal temperature expands rapidly setting up an acoustical shock wave. Usually experienced as a lightning bolt (plasma) and resulting thunderclap (shock wave).

Plasma shield
the ability of plasma to stop transmission of laser light.

Pockel's cell
an electro-optical crystal used to achieve a Q-switch.

Population inversion
a state in which a substance has been energized, or excited, so that more atoms or molecules are in a higher given excited state than in a lower resting state. This is a necessary prerequisite for lasing action.

Power
the rate of energy delivery expressed in watts (joules per second).

Power density
(irradiance) the amount of energy concentrated into a spot of particular size. It is expressed in watts per square centimeter and is the brightness of the spot.

Pulse
a discontinuous burst of laser as opposed to a continuous beam. A true pulse achieves higher peak powers than that attainable in a continuous wave output – usually pulsed in microseconds or shorter. (*See also* Gated pulse).

Spot size
the mathematical measurement of a focused laser spot. In a TEM_{00} beam it is the area that contains 86% of the incident power. This is the 'optical' spot size and does not necessarily indicate the size of the laser crater that will be made. The latter is the impact size.

Superpulse
an operating mode on the CO_2 laser describing a fast pulsing output (250–1000 times per second), with peak powers per pulse higher than the maximum attainable in the continu-

ous wave mode. Average powers of super-pulse (speed of tissue removal) are always lower than the maximum in continuous wave.

Thermal relaxation time

the rate at which a structure can conduct heat. When pulse times of a laser are shorter than the time required for heat to spread out of a target, the heat damage will be confined to that target.

Tunable dye laser

a laser using a jet of liquid dye, pumped by another laser or flashlamps, to produce various colors of light. The color of light may be tuned by adjusting optical tuning elements and/or changing the dye used. Common medical applications are with the 630 nm continuous wave red, and the pulsed 577 nm yellow and 504 nm green.

Q-switching

switching the 'quality' of a resonator, producing very high peak powers (millions of watts) but for very short bursts (nanoseconds) – usually achieved with a pockel's cell. This creates a 'sparking' and shock wave effect (*see* Photodisruption; Plasma and Mode-locking).

X-ray

a very short wavelength of light, producing ionization effects commonly associated with radiation hazards. Not a problem with surgical laser units.

1   Simplified physics

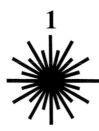

The word LASER is an acronym for Light Amplification by the Stimulated Emission of Radiation. Visible light is only one small portion of the electromagnetic spectrum (Fig. 1.1). Although the exact nature of light is still not understood, it does show properties both of discrete particles (photons) and waves. For the purposes of understanding the electromagnetic spectrum and lasers, we will primarily look at light in terms of its wave characteristics. A wave is characterized by four quantities: wavelength, amplitude, frequency and velocity (Fig. 1.2).

Properties of waves

Wavelength

The wavelength is the distance between two successive crests, or any other two points on the same parts of the wave. The color of

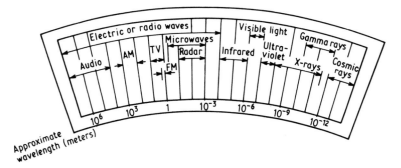

Fig. 1.1   The electromagnetic spectrum, showing the wide range of wavelengths from the very short cosmic rays to the very long radio waves.

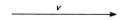

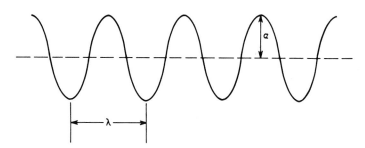

Fig. 1.2 The basic properties of a wave. The diagram shows a wave of velocity *v*, wavelength λ and amplitude *a*.

visible light is determined by its wavelength, which is measured in fractions of a meter known as nanometers (nm). One nm is equal to 10^{-9} meter. Our eyes are designed to see only a small portion of the spectrum. Visible light waves have wavelengths in the range of about 385–760 nm (Fig. 1.1). More energy is associated with shorter wavelengths (blue) than with longer wavelengths (red).

Amplitude

The amplitude is the height of the wave with maximum displacement from the zero position. Like an ocean wave, the greater the amplitude the more the power. In the case of light, the brighter the beam.

Velocity

The velocity of waves is a constant in a given medium, and is equal to about 186 000 miles/s or about 300 000 km/s, in a vacuum. The unique characteristic of light is that the speed of light is always constant in all frames of reference. In this sense its speed is an absolute and is further explained in Einstein's theory of relativity.

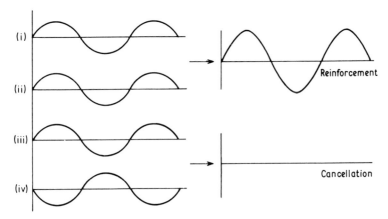

Fig. 1.3 Phase. Waves (i) and (ii) are 'in phase' since all the troughs and all the peaks are opposite each other. The result is a reinforced wave of double the amplitude. Waves (iii) and (iv) are 'out of phase' since troughs are opposite peaks. The result is a complete cancellation.

Frequency

The frequency is the number of waves passing a given point per second, and is expressed in cycles/s, or Hertz(Hz). The shorter the wavelength, the higher the frequency, since more waves will be able to pass a given point in a certain time.

Some further properties of waves also need to be considered.

Phase

Waves (of the same wavelength) are in phase when all the troughs and all the peaks are opposite each other. If two such waves meet, the result is a reinforced wave of double the amplitude (increased brightness). Conversely, if the waves are out of phase (troughs opposite peaks), then the result is a disappearance of the wave (Fig. 1.3).

Coherence and incoherence

Coherent light consists of unbroken waves that are of constant wavelength and have no phase differences either in time or space (temporal and spatial coherence). An analogy may be drawn between a pure musical note (coherent) and noise (incoherent).

Where does light come from?

Having considered some of the fundamental properties of light waves, we are now in a position to understand how a laser works, where its light comes from, and how this light differs fundamentally from ordinary light.

>   . . . and God said, 'LET THERE BE LIGHT!'. . .

In an atom, electrons are found to occupy certain discrete energy levels or orbits. These electrons are not free to have energies between levels or to take up positions between orbits, so that when the energy level of an atom is changed, the electrons must move up or down to the next orbital level. When an atom or molecule absorbs energy, electrons move into higher orbits, but fall back to their own less energetic resting orbits almost immediately. When an electron falls to a lower energy level, there escapes a tiny burst of surplus energy – a photon, the basic unit of light. The energy of the photon is simply the difference in energy between the two levels involved. Energy determines the wavelength, which is the color of the light. When many atoms in a medium undergo spontaneous orbital decay, the process is known as 'spontaneous emission' (Fig. 1.4). The decay of different atoms occur at random. Many energy levels are involved and the light is emitted out of phase and in all different directions. Different energy levels mean multiple colors of light, and all the colors combined produce white light. This is incoherent light, like taking the lid off a popcorn popper just as all the kernels begin to explode, the kernels of light travel in all different directions out of phase with one another.

A substance has the potential to become a lasing medium if it can have more atoms or molecules in a high energy state than in its resting energy state. This is known as a population inversion. Lasers are named after the medium that produces the light such as carbon dioxide, argon, etc. Different mediums emit characteristic colors of light which, in turn, are used for various medical applications. In most lasers, a medium of gas, liquid or crystal is energized (pumped) by a suitable source (light, electric discharge, radio frequency). The input of pumping energy raises electrons to higher energy levels in more atoms, more quickly than spontaneous decay can return them to their original level. Once there

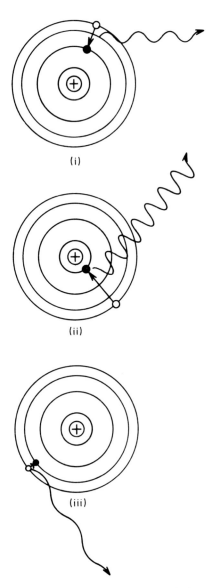

Fig. 1.4 Spontaneous emission. The diagrams show energized electrons in three atoms decaying back to their original orbits with the spontaneous production of a photon of light. The wavelength and amplitude of the emitted light varies according to the magnitude of the energy change, and is distributed in random directions. The overall light output is incoherent.

is a preponderance of these excited atoms (i.e. atoms having an electron in a higher energy level), a further process becomes probable in addition to the spontaneous emission just described. A photon from an initial spontaneous decay stimulates each excited atom in its path to emit an identical photon, in phase, of the same color and travelling the same direction. These photons actually fuse together in time and space producing a coherent output. This is known as stimulated emission, each photon stimulating another energized electron to produce a further photon, until a photon cascade of growing energy sweeps through the medium (Fig. 1.5) like a chain of dominoes falling. The waves of light produced in this way are reflected back and forth many times by mirrors at each end of the laser chambers (Fig. 1.6). These mirrors perfectly face each other and form a type of 'infinity tunnel' in which the light waves are trapped and bounce back and forth at the speed of light, increasing the amplitude (power or brightness) of the wave with each pass. In medical laser systems one of these mirrors is partially transmissive (like a two-way mirror) which allows the laser beam to leak out at this end. This is the beam of laser light. Some materials are more efficient at producing laser light than others, which is why the electrical voltage requirements vary so greatly for different lasers.

The laser beam is passed through some type of delivery system to the surgical field (Chapter 4). Flexible fibers or articulated arms deliver the beam, lenses focus and intensify the energy, or synthetic ceramic probes or sculpted fibers convert the laser output into heat in the tip of the device. The various attachments and delivery systems determine the utility of any laser.

Special properties of laser light

Laser light differs from ordinary light in much the same way that music differs from noise. Three particular properties are responsible for this difference: coherence, collimation and monochromaticity. Not all are important for surgical use, but an understanding of these concepts shows how future medical applications may develop.

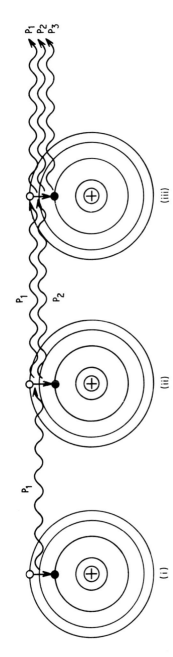

Fig. 1.5 Stimulated emission. The diagrams show energized electrons in three atoms in a substance which has undergone a population inversion, so that many of the atoms are in an excited state. The first atom (i) undergoes a spontaneous decay and emits a photon P_1. This interacts with a second energized atom (ii), and stimulates the emission of a second photon P_2 with precisely the same wave characteristics, and in perfect phase with P_1. Each of the identical photons can then further stimulate energized atoms to produce additional identical photons, as P_3 etc.

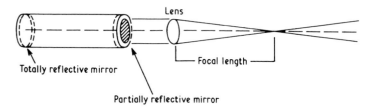

Fig. 1.6 Schematic diagram of laser chamber. The growing cascade of light produced by the stimulated emission is reflected back and forth between the mirrors at either end of the laser chamber until the beam leaves the chamber through the partially reflective mirror. It can then be focused by a lens, and pass into a suitable delivery system (Chapter 4).

Coherence

Ordinary light, from a lamp or fire, is incoherent, and consists of light waves radiating (shining) in all directions out of phase with one another. Since multiple wavelengths (colors) are produced, it is not possible to phase the waves. When a handful of stones is thrown into a pool of water, the choppy wave pattern created on the surface is incoherent because the ripples produced by each individual stone are out of phase with one another.

Laser light is coherent. Its light is also of one wavelength (color) which allows the waves to synchronize when they are phased together. When a piano tuner strikes his tuning fork, he phases the piano string to the tuning fork. The surf breaking onto the beach is also a phased wave pattern. When the peaks and troughs align, they are in phase.

Combined with the pure color of light, the coherence allows the creation of holograms – three-dimensional images created by the laser. Not only unique as an art form, the use of holography may change the practice of medicine. These include three-dimensional X-rays; images that 'float' over a patient while the surgeon works; endoscopic holography for pathological diagnosis; stress test scanning in orthopedics and other areas for mechanical stress imaging.

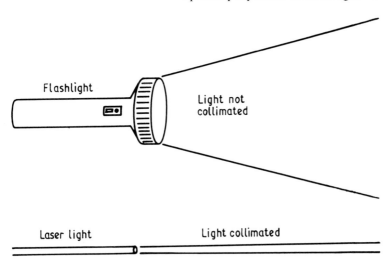

Fig. 1.7 Collimation. The light from a flashlight (torch) is not collimated and will spread out along the length of the beam. In contrast, the light from a laser is highly collimated, and has an insignificant spread along the length of the beam.

Collimation

A collimated beam is parallel and does not diverge, in contrast to the light from a flashlight (torch) which will spread out as it travels further and further away (Fig. 1.7). Laser light is for practical purposes parallel, so that a laser pulse fired at the moon produces a spot half a mile wide, at a distance of 240 000 miles. For everyday purposes this spread is insignificant. This means two things in medicine – there is a minimum loss of power along the beam, and the beam can be focused to intensify its effect or couple it into a slender single fiber. A laser beam can be a billion times brighter than sunlight.

This characteristic is what really allows lasers to be useful for most surgical applications. It allows the majority of the energy generated to be collected and then intensified into tiny spots.

Monochromaticity

Monochromaticity indicates that the light is all of the same wavelength (color). Ordinary light sources produce light from a 'hot body' process. Light from the glowing filament in a lamp usually consists of a mixture of all possible colors in a broad range. This results in white light. All the light of the laser is concentrated in a few discrete wavelengths (frequently one). Lasers produce pure colors of light. Different materials emit characteristic colors (Table 1.1).

Table 1.1 Characteristic wavelengths of light from some lasers

Laser	Color	Wavelength (nm)
Carbon dioxide	Infra-red	10 600
Argon	Blue	488
	Green	515
Nd:Yag	Infra-red	1064
Ho:Yag	Infra-red	2060
Er:Yag	Infra-red	2940
KTP	Green	532
Krypton	Red	647
	Yellow	568
	Green	531
Ruby	Deep red	694
Helium neon	Red	632
Gold vapor	Red	632
Copper vapor	Yellow	578
Dye laser	(variable with dyes)	
	Red	632
	Green	504
	Yellow	577
Excimers:	Ultraviolet:	
Argon fluoride		193
Krypton fluoride		248
Xenon chloride		308
Xenon chloride		351

Color is not the most important characteristic for surgical use, but other applications are developing where the pure colors are critical. One such area is photodynamic therapy discussed in Chapter 5.

The laser medium

A laser is usually named after its active medium, the substance which exhibits the lasing action. This can be a liquid, solid or gas. Substances which have been used include the following:

Solids (Crystals): Ruby, Nd:Yag (Yttrium Aluminum Garnet, doped with neodymium – a rare earth element), Nd:Glass, Er:Yag (Erbium Yag), alexandrite, Nd:GASG (Gadolinium Aluminum Scandium Garnet doped with neodymium) or Holmium:Yag (actually Chromium-Thulium-Holmium:Yag).

Gases: Carbon dioxide, argon, krypton, helium–neon, carbon monoxide, hydrogen fluoride, argon fluoride, xenon chloride.

Liquids: Dyes of different types which allow emission of various wavelengths depending on the dye, giving rise to the name Tunable Dye laser.

Other substances have also been used, but the lasers which have found the widest applications in medicine at the moment are those based on carbon dioxide, argon, Nd:Yag, krypton, Ho:Yag and various dyes. All of these substances need to be energized into a state of population inversion. Solid and liquid lasers tend to be energized optically, usually by flash lamps or another laser, and gas lasers tend to be energized electrically by direct current, or radio frequency (RF).

What are sealed tube lasers? (Carbon dioxide)

When the carbon dioxide gas mixture is energized and stimulated to emit its light, there is a disassociation of the molecule into carbon monoxide and a free oxygen radical, unlike argon or neodymium atoms which do not break apart (Fig. 1.8). The resulting molecule is no longer able to produce light. Complicating this is the fact that the electrodes in the laser tube give off contaminants which further degrade the gas mixture. In medicine, three basic configurations of carbon dioxide lasers have resulted.

Flowing gas systems

Because of the disassociation of the molecule and electrode contamination, a flowing gas system is used to purge the tube and replenish the gas. This requires cylinders of replacement gas,

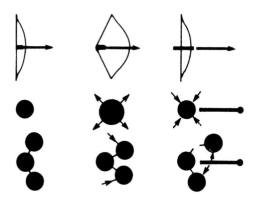

Fig. 1.8   Light emission from a carbon dioxide gas mixture.

pressure regulators and a vacuum pump to draw the gas through the system. All of this adds to the maintenance, noise and expense of operating the laser. On the other hand, flowing gas systems produce a reliable, steady output, and easily generate high powers (for medicine) of 50–100 W. They also produce stable, low power outputs of around 1 W.

Some manufacturers have developed a microflow system. These still utilize a flowing mixture of gases, but at significantly lower flow rates than the previously designed systems.

Free space, sealed tube lasers

These are a newer generation laser tube that eliminates the need for replacement laser gas mixture, regulators and vacuum pumps. They can produce power levels comparable to flowing gas systems up to about 100 W.

This sealed tube laser uses direct current (DC) excitation as do the flowing gas systems. The electrodes are specially treated, and catalysts and inhibitors are added to the self-contained gas mixture to retard breakdown of the gas. This allows for a sealed-tube type of laser system. After several years the tubes need to be recharged as the power begins to decrease gradually from the high end. Gold vapor is sometimes added to these tubes to retard breakdown of the electrodes and increase the longevity and power of the tube.

Radio frequency (RF) waveguides

These are also sealed-tube systems but instead of using direct current applied through electrodes to energize the gas they utilize a radio frequency that is transmitted transversely across the tube to excite the gas molecules, which eliminates the electrode problems. The ceramic tube of the laser is also impregnated with catalysts and inhibitors to retard breakdown of the gas. The radio frequency is contained within a waveguide structure of the laser. There is no interference with monitors or electrical equipment. RF waveguide lasers historically produced lower power of 25 W and less. Newer technology enables higher power outputs of 100 W or more, as well as 100 W average power superpulses, which is not possible with the other types of lasers.

Sealed-tube systems of both varieties are generally simpler to operate, quieter and require less routine maintenance than flowing gas systems. Both flowing gas, and sealed-tube systems each have their own advantages.

Energy concepts (Watts, power density, joules, fluence)

Power is simply a measure of the rate of energy delivery in joules/second (J/s) and is expressed in watts (W).

Of greater importance is the amount of power which can be focused into a spot. This power density or irradiance (sometimes called spot brightness), of a laser is the number of W/cm^2 of a spot, and is the single most important factor in the effective applications of a laser. The surface area of the spot (spot size controlled by the surgeon in the field), and the total power in watts (set at the laser by the operator) determine the power density as follows:

$$\frac{Watts \times 100}{\pi \times r^2} = W/cm^2$$

Power density over a spot determines the rate of tissue removal within that spot. One can effectively change the size of the paintbrush (spot size) with which you are working without changing the overall rate of tissue removal (power density) within that spot by varying the spot size and power. The larger the spot, the

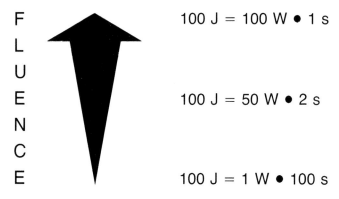

F
L
U
E
N
C
E

100 J = 100 W • 1 s

100 J = 50 W • 2 s

100 J = 1 W • 100 s

Fig. 1.9 Fluence combines the concepts of power density with total delivered energy, expressed in J/cm^2. Increasing the fluence limits lateral thermal damage. Flux is a similar concept but ignores the parameter of power density.

greater the power required to maintain the same power density, shown as follows:

	0.6 mm spot	2.0 mm spot
1900 W/cm^2:	10 W	60 W

The total energy within the beam is expressed in joules (J). Power multiplied by delivery time equals the number of joules: 10 W delivered for 3 s is 30 J of energy. Joules indicate the total energy delivered but do not indicate how concentrated this dose of light is.

Fluence combines the concepts of power density (spot brightness) and dosage (joules), and is expressed in joules/cm^2 (Fig.1.9).

Optical concepts

The focal length of the laser lens

On non-fiberoptic delivery systems (CO_2 lasers), lenses are interchangeable. The smaller the focal length of the lens, the smaller the spot size and the greater the power density at any given power. CO_2 lasers are able to achieve spot sizes in the range 0.025–0.05 mm, but most medical systems produces spots of 0.1–0.8 mm.

The wavelength of the laser

Each laser has its own characteristic wavelength (Table 1.1). The shorter the wavelength, the smaller the spot which can be produced. An argon (515 nm) or KTP (potassium titanyl phosphate, 532 nm) would be able to produce spots much smaller than a CO_2 laser (10 600 nm). However, a type of laser would be chosen for its specific effects on tissue rather than its spot size capabilities. On any type of fiberoptically delivered laser, the smallest focused

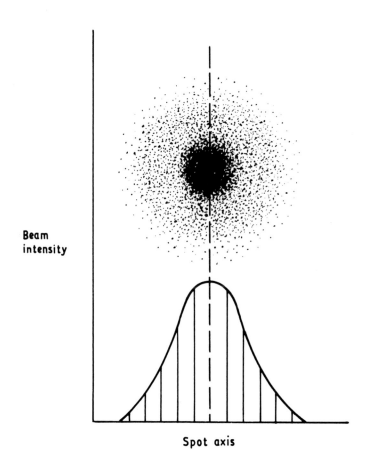

Beam intensity

Spot axis

Fig. 1.10   Mode TEM_{00}, with most of the power at the center of the spot.

spot which can be achieved is the size of the tip of the fiber, regardless of the wavelength. An argon laser delivered through a 0.4 mm fiber could not be focused any smaller than a 0.4 mm spot.

The mode

Mode refers to the distribution of power over the spot area, and determines the precision of the operative spot size. The technical term is transverse electromagnetic mode (TEM) of the beam. The most fundamental mode, known as TEM_{00}, shows a power distribution over the spot which has most of the power at the center. A graph of beam intensity against its axis is shown in Fig. 1.10. This mode can be focused to the smallest spots and is

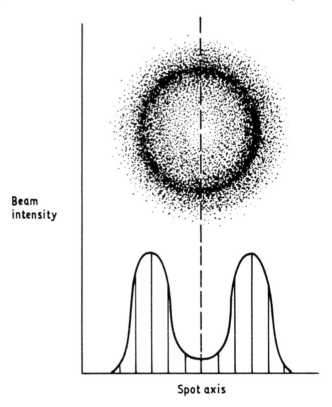

Beam intensity

Spot axis

Fig. 1.11   Mode TEM_{01}, with a cold region in the center of the spot.

the cleanest beam. This concept is not applicable to fiberoptically delivered lasers since the fiber destroys the mode structure. Most of the medical CO_2 lasers produce this clean beam.

When the intensity is not distributed in this fundamental mode, it is said to be in a multimode distribution. Many different mode structures can be present simultaneously. A common mode which occurs in the first fraction of a second of a pulse for high powered (100 W) CO_2 lasers is known as TEM_{01}, which exhibits a cold region in the center of a doughnut shaped burn (Fig. 1.11).

Q-switching and mode-locking

Q-switching is a term used to describe the switching of the quality of a resonator, causing it to produce high peak powers but lasting for very brief intervals, frequently in nanoseconds (10^{-9} s). This is usually associated with ophthalmic Nd:Yag lasers, causing a spark and snap in the eye. However, some other types of lasers in medicine are beginning to employ this pulsing technique, holding just short of the spark and snap. Mode-locking is also used in some ophthalmic Nd:Yag lasers, creating similar end results but through a different mechanism, resulting in a train of much shorter pulse widths, in the range of picoseconds (10^{-12} s).

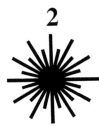

2 Laser – tissue effects

Heating of tissue

The effects created by surgical lasers, namely cutting, vaporizing and coagulation, are all caused by heating of the tissue. Table 2.1 describes the effects on soft tissue as temperatures increase – from laser or any other source of heat. The wavelength (color) of the laser is important only in determining how efficiently this heat transfer occurs, and over what volume of tissue.

Table 2.1 Absorptive heating

Temperature °C	Visual change	Biological change
100	_____ Smoke plume	Vaporization, carbonization
90–100	_____ Puckering _____	Drying
65–90	_____ White/Grey _____	Protein denaturization
60–65	___ Blanching _____	Coagulation
37–60	None _____	Warming, welding

Lasers offer advantages over other heat sources because of precision, surgical access, or selectivity. One unique characteristic of lasers is that during vaporization they can mold and sculpt tissue. Although frequently advantageous, lasers are not always the only surgical alternative to cutting or coagulation of tissue.

If tissue temperatures significantly exceed the boiling point of water at the point of impact, tissue will explode and emit steam and particles. At a small point an incision is created; at larger spot sizes tissue debulking occurs. At lower temperatures protein coagulation and desiccation may occur. It is useful to review the mechanisms of tissue heating employed by laser, and even electrosurgery.

Heat is not energy, but only a mechanism of the transfer of energy. It results from the continuous motion and vibration of atoms and molecules. The energy of the laser beam is transferred into tissue as heat, because the light frequencies cause the tissue molecules to vibrate at a rate at which absorption occurs. Heat can be measured quantitatively, with units of measure typically either calories or British Thermal Units (BTU).

Temperature measures the intensity, but not the quantity of heat. For instance, a 50 gallon drum of water at 72 °F contains significantly more heat than one ounce of water at 100 °F, as shown by the ability of the 50 gallons to melt more ice than the one ounce at 100 °F. This will be seen with the spark created by a Q-switched Yag laser to 'cold cut' internal eye membranes. The actual temperature internal to the microscopic spark is several thousand degrees centigrade, but it is so microscopic that it contains little heat – hence the term 'cold cutting'.

Another way to look at this is that a bulkier object will contain more heat than a smaller object at the same temperature. A 1000 μm tapered fiber used as a hot knife can contain more heat than a 600 μm tapered fiber because it has more bulk to it. This creates faster cutting through denser material because of the increased heat in the bigger fiber. It also makes for more mechanical stability.

Absorption of light by tissue

The nature of interaction of all laser light with biological tissue can be described in terms of reflection, transmission, scattering or absorption (Fig. 2.1).

In order for light to heat tissue it must be absorbed. If it is reflected from or transmitted through tissue there will be no effect. If the light is scattered, it will be absorbed over a large volume and its effects will be more diffuse.

In some ways these concepts are intuitive and relate to the energy concepts discussed in Chapter 1. An analogy is the use of a magnifying glass to focus sunlight and create a burn. The brighter the sun (power) the more it burns. The smaller the spot created

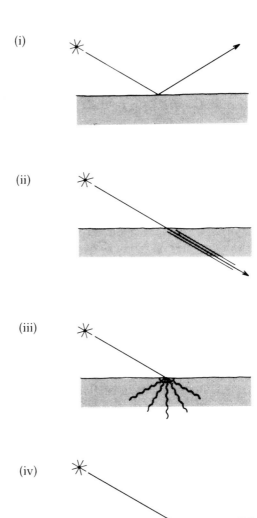

Fig. 2.1   (i) Reflection. A laser beam is reflected from the surface of a tissue
                 and has no effect.
           (ii) Transmission. A laser beam is transmitted through a tissue and has
                 no, or only minimal, effect.
          (iii) Scattering. A laser beam is scattered by a tissue and absorbed over
                 a large area. Its effects are diffuse and weakened.
           (iv) Absorption. A laser beam is absorbed by a small volume of tissue
                 and exerts its effects within this volume.

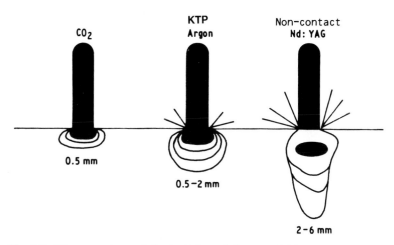

Fig. 2.2   Tissue reaction and depth of thermal damage of CO_2, argon and non-contact Nd:Yag lasers.

with the lens (power density) the hotter it burns. The longer the sun is focused in one place (total energy) the more extensively it burns. The wetter the target (vascularity) the longer it takes to create the same effect.

Figure 2.2 compares the relative effects of CO_2, argon (or KTP), and Nd:Yag lasers. This shows the relative spread of thermal damage from the raw beam and ignores other very important factors such as delivery systems, power densities, powers and pulse times.

As indicated, the CO_2 laser is the most highly absorbed, only 30–90 μm in water, and causes the most intense heating effects – cutting and vaporization. It will coagulate tissue, but only superficially.

The pulsed, Ho:Yag laser provides the next most highly absorbed wavelength of laser, and is able to be transmitted through fluid and conventional fibers.

The Nd:Yag laser scatters its light over a broad volume of tissue causing coagulation as deep as 4–6 mm if desired. At higher power densities it will vaporize, but leaves an underlying deep coagulation beneath the base of the crater. In Chapter 4 we will examine how sapphire probes, or sharpened or sculpted fibers can completely change these characteristics by making the laser a hot surgical knife.

The argon (488 and 515 nm) and the KTP (532 nm) lasers produce very similar if not identical tissue effects from a clinical perspective. They are generally considered superficial coagulators, coagulating 0.5–2.0 mm deep. At smaller spots and higher power densities they can also be used for finer cutting. For vaporization and debulking they proceed very slowly because of these laser's relatively low power of around 15 W. There are some differences in absorption peaks of argon and KTP, as well as the high frequency pulsing of the KTP, but for a general understanding of tissue effects they are considered equivalent.

Vaporization of tissue occurs when cellular fluid is heated to its boiling point. The rapid rise in intracellular temperature and pressure causes an explosion of the cell, throwing off steam and cellular debris known as laser plume or smoke (Fig. 2.3).

Soft tissue vaporizes at 100 °C. If the remnant carbon is allowed to collect, and lasing is attempted through this carbon, the lasing point will begin to incandesce to an orange glow as tissue heats to over 1500 °C. This would cause extensive thermal damage resulting in a branding iron type burn and concomitant scarring. It is usually created by lasing at power densities that are too low for excessive lengths of time.

Cutting is simply vaporizing tissue along a line, using very small spot sizes of 0.1–0.5 mm. Defocusing the beam slightly, using higher power to maintain power density, and cutting with this broader beam, results in less precision but better hemostasis for vascular areas.

Hemostasis

Coagulation of tissue with the laser has a broader meaning than simply achieving hemostasis. The laser has the ability to destroy cells by denaturing the protein without necessarily vaporizing it. This would be like heating egg white and seeing it turn white as it coagulates. Deep coagulation may sometimes be desirable as when treating bladder tumors. The non-contact Nd:Yag laser has the deepest coagulating abilities, followed by the argon. Hemostasis or cautery, is achieved by sealing the ends of the capillaries or vessels with the laser heat secondary to cutting or vaporizing. The greater the blood flow through a vessel, the more it acts as a heat sink, and the greater the difficulty in achieving hemostasis.

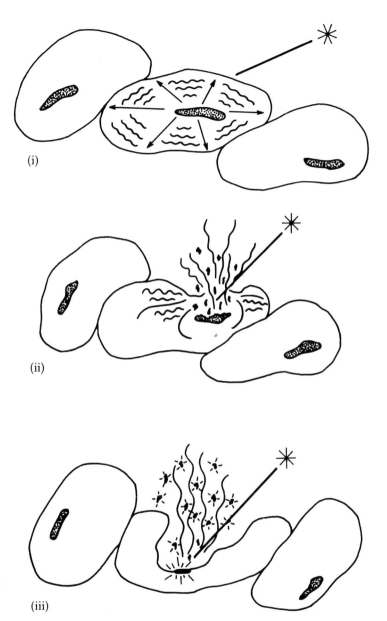

Fig. 2.3   (i) A cell has absorbed laser light and is heated to boiling point. The cell is destroyed.

      (ii) The cell explodes, throwing off steam and cellular debris.

     (iii) The steam and debris rise from the site of impact and are carbonized in the laser beam.

Applying pressure, when possible, to slow the blood flow helps to achieve better hemostasis.

Hemostasis achieved with a laser may be similar to spray coagulation achieved with electrosurgery, with the laser being more controllable and predictable, and able to work in a non-contact mode. The argon beam coagulator, which is a monopolar electrosurgical probe, is larger than a laser fiber and also works in a non-contact mode. Lasers do not always eliminate the need for electrosurgery and in certain cases the advantages of laser over electrosurgical coagulation have been overstated. Electrosurgery may take many different forms, from delicate coagulation with bipolar forceps, fine cutting with a micro-needle, to the charring and burning in the spray coagulation mode, or the superficial to very deep effects of the argon beam coagulator which is not to be confused with a laser. Lasers cannot perform tissue excavation or snaring as can electrosurgery. These modalities complement one another.

However, electrosurgery has many major disadvantages, which include burns, interference with monitoring equipment and pacemakers, and unrecognized energy (stray current) leading to unrecognized tissue injury. The cause of 'stray current' includes insulation breaks in electrodes and capacitive or direct coupling between the active electrode or other metal instruments. Capacitive coupling poses the greatest risk for injury when using monopolar electrosurgery in laparascopic surgery. None of these electrical problems occur with lasers.

Hard tissues

Since bone and cartilage contain relatively little water, they vaporize differently from soft tissue. Bone has a tendency to heat as does carbon, and conducts heat into the adjacent soft tissue. To protect these tissues and limit the thermal damage, a superpulse mode is preferred with the CO_2 lasers. This appears as a continuous beam, but is actually cycling on and off from 250 to 1000 times per second with high peak powers (250–500 W) on each spike. This allows a cooler cutting of the bone or cartilage. Superpulse is useful but commercial claims of varying different superpulse parameters is exaggerated. Bone may also need to be continuously irrigated to prevent flaming. Cutting of bone with lasers is not entirely satisfactory but investigators are examining the use of the 193 nm argon fluoride (excimer) laser and the 2900 nanometer Er:Yag and hydrogen fluoride lasers because of their non-thermal type of cut.

Pulsed laser energy as a means to refine tissue effect

Advances over the last several years include the recognition of the laser as a potentially destructive heat source if used for excessive periods of time or at too low a power density.

A true pulse of energy can deliver enough energy to create the desired effect, but at such a rapid rate that no lateral heat conduction can occur. This type of pulsing concept is something entirely different from simply setting a timer (i.e. 0.05 s) on the laser to make the emission short, though this is usually labeled as a 'pulse' on the control panel. A true laser pulse compresses the energy delivery so that peak powers go much higher than they otherwise might in a steady-state emission, and the pulse durations are typically in the microsecond range or less.

From a conceptual viewpoint, one can look at the absorption volume of the light into tissue. Light is instantly absorbed by a certain volume of tissue, depending on absorption coefficients, tissue color, etc. Before any laser damage can occur outside this tiny absorption volume the tissue within that volume must be heated to in excess of 100 °C. Tissue then vaporizes and the laser beam moves onto the next block of tissue.

All materials, including tissue, conduct heat at various rates. While the laser-induced heating is occurring within the volume of tissue, the heat that is building up begins to conduct to adjacent tissue outside of this laser volume. This generates (sometimes significant) lateral heat damage to adjacent tissue. Even though heat does conduct rapidly in tissue, it does take a definite time before it can conduct outside the absorption volume. Arbitrarily, for the sake of discussion, heat begins to conduct in the 650 μs range.

A race is now occurring between the time it takes to vaporize one tiny volume of tissue, and the time required for heat conduction to begin. If the laser can vaporize the volume in less than 650 μs then all the heat which was generated has been confined to the targeted, and eliminated tissue. If the laser takes longer than 650 μs to vaporize that volume, then lateral heat conduction begins to occur.

Power is the rate of energy delivery of the laser, and most lasers which are continuous wave (cw) output lasers cannot deliver the energy fast enough to beat the race with lateral conduction. Pulsed lasers, however, can compress their energy delivery to deliver very high peak power (rate of energy delivery) pulses which can beat the vaporization versus lateral heat conduction rate.

Pulsed lasers, such as superpulse on the CO_2 laser, pulsed Ho:Yag or, to a lesser extent, the pulsing of the KTP laser, vaporize or incise tissue very precisely because of the limitation of heat conduction. Pulsed lasers such as the yellow pulsed dye can limit heat generation to only the targeted capillaries and spare normal skin, to prevent scarring.

Other, non-heating, laser tissue effects

Other tissue effects besides heating become possible when pulsing laser energy high enough. These are sometimes described as non-linear laser effects, since the effects are unrelated to simple linear rises in temperature.

Once the intensity of the light is high enough, generally created by pulsing, the actual electric field in the focused spot will induce an ionization of material within the target spot. This is the creation of small plasma sparks, which produce associated snaps, or tiny sonic booms. Varying somewhat in intensity, and actual mechanism, these effects are utilized to fragment biliary and kidney stones with pulsed dye lasers, or to cut membranes in the eye with a Q-switched ophthalmic Yag laser.

Other lasers, such as the 193 nm argon fluoride laser, produce a photolytic type of effect on biological tissue. The energy contained within the ultraviolent photons is so high that it photochemically causes a disassociation of carbon bonds in biological tissue. This clean and cool laser effect is used in ophthalmic applications to resurface the cornea.

3

Properties of individual lasers

The carbon dioxide (CO_2) laser

As will be discussed later, the Nd:Yag laser, when used with contact probes, can achieve similar effects to the CO_2 laser in terms of cutting or vaporizing. Though more difficult in its operation to learn, the CO_2 laser affords speed and versatility in terms of depths and angles of approach to an experienced user. Its unique strengths lie with the use of microscopes for a no-touch method of cutting or vaporizing tissue as performed with craniotomy, microlaryngoscopy or colposcopy.

When used for cutting, the beam is used in focus, at its smallest spot. The depth of the cut is determined by power density and the speed of the incision. The slower the stroke or higher the power density, the deeper will be the cut. To achieve good clean edges to the incision, with minimal charring, traction across the incision line is essential. A mode such as superpulse will give a cleaner incision while slightly decreasing hemostasis. The general rule is to set the spot at its smallest, then the speed and physician comfort level of the incision is controlled directly by the power which is started at moderate settings and increased as needed.

Incisions made with the CO_2 laser heal in much the same way as a conventional wound histologically, though the order of the healing process is slightly different. Scars and wound strength are identical at 20–30 days. In practice, most surgeons see no clinical difference after 7–10 days.

Vaporization of tissue may be performed with a focused or defocused beam (Fig. 3.1). Although small areas can be vaporized with a focused spot, larger areas are better handled with a defocused (broader) spot, and a higher power applied to compensate

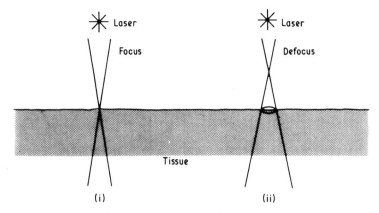

Fig. 3.1 Focus and defocus. In (i) the laser beam is focused on the tissue in a small spot. In (ii) the laser beam is focused in front of the tissue, so that a defocused spot, of larger area and lower power, impinges on the tissue.

for the dilution of power density. Pulsing the laser in short bursts allows one to use the maximum power output of the laser to advantage, without concern for excessive damage.

The general rule is first to select the spot size with which you want to vaporize, then to increase power to a suitable power density, as reflected in the geometry of the impact crater. If spot sizes of 2.5 mm or larger are acceptable, then increase the power of the laser to its maximum (assuming no more than 100 W lasers), and set the spot to deliver a suitable power density as reflected in the geometry of the impact crater.

Even though lasing through char is a classic mistake in using the CO_2 laser, one can very precisely remove skin lesions a layer at a time by painting an even layer of black char over the skin, then wiping away this layer with a wet sponge before relasing. By not lasing through char high temperatures are avoided but it does destroy a very superficial layer of skin that may be wiped away. The cosmetic results can be good. When the perfect smoothness of the healed site is not a consideration, more of a stripping technique using higher power densities and faster strokes, with blunt spot sizes, will result in precise, superficial vaporization.

The argon laser

The first medical use of a laser was the argon laser in the treatment of diabetic retinopathy in 1965. Since that time, extensive experience with argon lasers has resulted in this photocoagulator becoming the treatment of choice for this retinal disorder. Dermatology is another major user of this color of laser. The laser operates as a sealed tube system with an operating life of only a few years. It is possible to rebuild and recharge these tubes, rather than replacing them with an entirely new tube, at significant savings to the user.

Argon lasers produce a visible blue–green light (488 and 515 nm) which is easily transmitted through clear aqueous tissue. Certain tissue pigment (red or black) such as melanin or hemoglobin will absorb the light very effectively. This principle of selective absorption is used to photocoagulate pigmented lesions such as portwine stains (on skin) or endometriosis (intra-abdominally).

This light passes through overlying skin, without significant absorption, and reaches the pigmented layer of portwine stains (or other colored lesions) to effect capillary desiccation and protein coagulation. Gradual blanching then begins to occur, sometimes taking many months for the stain to fade completely.

When the beam is focused to a very small spot and power increased, the power density is high enough to result in cutting or vaporization of tissue. Smaller diameter fibers result in high power densities.

Many makes and models of ophthalmic lasers are available although its use in general surgery is more limited. Several different models with various power outputs are available from manufacturers. Fiberoptic delivery of the beam is one of the primary advantages of this laser in ophthalmology, dermatology and gynecology. It is a continuous wave laser. When using it to 'pump' dye lasers, the resulting dye laser output will also be a continuous wave.

Frequency doubled Yag including KTP (potassium titanyl phosphate) laser and other 532 nm lasers

This is a Nd:Yag laser that uses a crystal of KTP (or alternatively other crystals such as KDP) at its output to change the color of light from that of the Nd:Yag (1060 nm) to that of the KTP or KDP wavelength (532 nm green light). In industry this is commonly referred to as a frequency doubled Yag laser. The only difference here is that all of the residual 1060 nm light has been filtered out to leave only 532 nm. It has nothing in common with the tissue effects of a Nd:Yag laser.

The wavelength is slightly different from that of the argon laser in that it is a greener color. This wavelength has a higher specificity for hemoglobin in skin lesions and endometriosis, although clinically their effects are identical (see argon laser tissue effects).

The argon laser produces a continuous wave output of a steady power, where the KTP is a high frequency Q-switched system of around 25 kHz. Both produce an average maximum power of around 15 W, and both are fiberoptically delivered. The advantages and disadvantages of each type of system come from comparing the operation of the laser itself, ease of use, types of fiber delivery systems, and safety features. Both types of laser have many medical uses. Fiberoptic delivery of argon, KTP and Nd:Yag is through flexible fibers which are used in contact, near contact or non-contact techniques (Fig. 3.2).

The Nd:Yag (neodymium:yttrium aluminum garnet) laser

The laser that is best suited for primary coagulative properties is the Nd:Yag (Fig. 3.3). When used as a non-contact beam, it coagulates 2–6 mm in depth and with some manipulation will handle vessels up to around 4 mm.

Nd:Yag is a solid crystal which is stimulated to emit in the near infrared light at 1060 nm. Medical units produce 25–120 W of maximum power and the laser energy is fiberoptically transmitted.

The light is transmitted through clear liquids, which allows its use in the eye or other water-filled cavities such as the bladder or uterus. Its absorption by tissue is not as color specific as argon, but the darker the tissue, the better the absorption. It is an excellent tool for tissue coagulation in endoscopic applications such as bronchoscopy, cystoscopy and gastroscopy.

Its use with contact probes made from sapphire, diamond or quartz allows both selective and fine cutting, vaporization, or coagulation of tissue with precision and avoidance of excessive tissue damage. Synthetic sapphire probes and other contact cer-

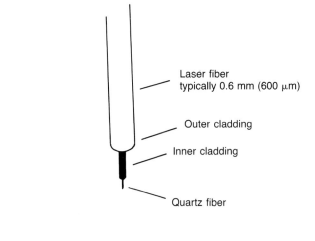

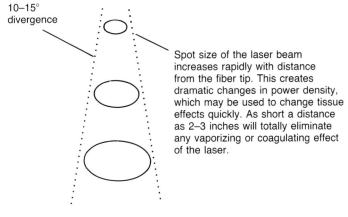

Fig. 3.2   Laser fiber and beam divergence.

amic materials are commonly used for medical applications (Chapter 4). This creates an overlap in tissue effects more commonly associated with the CO_2 laser.

Holmium:Yag laser

This laser, like the Nd:Yag, is also a solid-state system consisting of a crystal. It actually comprises several elements including chromium, thulium, holmium:yttrium aluminum garnet. Its mid-infra-

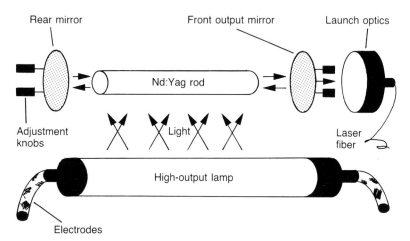

Fig. 3.3 Nd:Yag laser head. Like the CO_2 laser (Fig.1.6) the Nd:Yag uses a mirror at each end of the cavity to amplify the light. The Nd:Yag crystal rod does not conduct electricity, however, so that a high output flash lamp is enclosed in the head of the laser as the rod's 'energy source'. The emitted beam is focused by the launch optics into the laser fiber.

red wavelength of around 2.1 μm (2.06) is highly absorbed by fluid, yet it is still able to transmit through a flexible quartz fiber. It is also a pulsed laser, thus limiting thermal damage to a very small area.

Its highly absorbed wavelength and high energy pulses limit the spread of heat, making the soft tissue effects similar to those of a CO_2 laser. The lateral border of thermal damage may be slightly larger than that created by a CO_2 laser.

Its ability to transmit through a regular fiberoptic, and work in irrigation fluid, gives it the convenience and ease of use of other fiberoptically delivered lasers. Current primary medical uses are for cardiovascular recannalization, and for arthroscopy in orthopedics but other applications are developing.

The dye laser

Several wavelengths of dye lasers are now in medical use. In this laser a suitable organic dye is illuminated with a strong light source, usually the beam of an argon laser or flash lamps, to produce light of various wavelengths. The color of the laser can be controlled by varying the type of dye and the tuning elements in the laser – hence the name tunable dye lasers. Dye lasers are generally more mechanically unstable with breakdown problems and are treated appropriately.

These lasers are used where the selective absorption characteristics of the tissue suggest the application of certain colors of laser light. The tunability of this laser is used in ophthalmology to perform surgery in different areas of the retina with selectivity based on the various colors.

A 577–585 nm (yellow light) pulsed dye laser is used to great advantage in cosmetic vascular lesions because of the high specificity of hemoglobin to this wavelength. The high peak power, short duration pulse also helps to limit thermal damage by applying the beam at a faster rate than heat damage can spread to adjacent tissue. This pulsing technique is important in several laser applications.

A 504 nm pulsed dye laser is used to fragment kidney stones impacted in the ureter or otherwise inaccessible to shock wave lithotripsy. This laser sets up a concussive shock wave on the stone when the fiber is fired in direct contact with it causing stone fragmentation. Similar dye lasers are used for gallstones found in the biliary tract.

A continuous wave dye laser, at 630 nm (red light), is used in photodynamic therapy (Chapter 5). The red light corresponds to a particular absorption peak of a drug such as HpD (hematoporphyrin derivative). The output is pulsed when driven by pulsed power supplies (like pulsed flashlamps), or continuous wave when driven by those types of lasers (like the argon).

The copper vapor laser

This is a pulsed laser producing both yellow (577 nm) and green (515 nm) light. It is used for dermatologic applications similar to the yellow pulsed dye applications. This laser makes the shortest pulses of any of the yellow dermatologic lasers – lasting only nanoseconds. It may be focused to very small, 100 μm, spots to individually trace capillaries in lesions.

The excimer lasers

Excimer is a whole class of a dozen or so lasers. The term 'excimer' is derived from excited dimer and refers to a molecule which is comparatively stable when excited, but, when it loses energy by emitting a photon, splits up into its component parts. Conditions are therefore ideal for lasing, since this removal of molecules in the resting state results in a population inversion. Characteristically, excimer lasers emit in the ultraviolet spectrum, and deliver energies of about 0.1 J in a pulse duration of around 20 ns (10^{-9} s). Three primary types of excimers are in use at present in medicine.

The 193 nm argon fluoride excimer wavelength is transmitted via an articulated arm (like the CO_2 laser). Special types of fibers are being developed. This wavelength appears to remove tissue by a non-thermal mechanism of photolysis. This results in a high degree of precision, creating incisions on the cornea with no thermal shrinkage, and achieving predictable depths per pulse in the micrometer range. It is used primarily for corneal sculpting, and other applications include dentistry and orthopedics.

The krypton fluoride laser produces ultraviolet light of 248 nm. Due to the potential mutagenicity of this wavelength its medical use was limited. It is however being used in cardiovascular recannalization. The xenon chloride laser, producing light at 308 nm, is easily transmitted through conventional fibers and is not mutagenic. It still contains a high energy per pulse, and is aggressively being evaluated for the cardiovascular applications of laser coronary angioplasty. Many of the clean tissue effects which are ascribed to the wavelength of the excimer lasers are usually due to the natural pulsing of this class of laser.

The ruby laser

Invented in 1960 by T. Maiman, ruby lasers were largely ignored as a medical tool. Initially used for various retinal disorders, they were replaced by argon lasers in the 1970's.

Ruby is the common name for a sapphire host crystal doped with small amounts of chromium ions. The active medium is a ruby crystal in the form of a cylindrical laser rod. Continuous wave versions are possible but impractical for most situations.

The Q-switched ruby laser, which emits at the red wavelength of 694 nm, resurfaced in the 1980's for the treatment of tattoos and benign pigmented skin lesions.

The semiconductor diode lasers

Like transistors and related electronic components, diode lasers are solid-state devices made out of semiconductor crystals. Diode lasers are similar to the light-emitting diode (LED) display components used in many electronic appliances, in that they emit light when an electric current is passed through them. However, laser light is emitted that is much more directional and monochromatic than the light emitted by an LED, that can be focused to the very small spot sizes needed to do precise laser surgery.

Diode lasers are usually gallium aluminum arsenide (GaAlAs) semiconductor devices that emit laser energy at a nominal wavelength of approximately 800 nm, with continuous output powers of up to 10 watts or more. Ophthalmic diode lasers are used for the photocoagulation of retinal and other ocular tissue, as an alternative to the argon laser for some procedures. Their compact size and high laser output efficiency offer significant ergonomic and economic advantages, making them very attractive for office-based procedures. Numerous other therapeutic uses of diode lasers are being investigated.

New diode laser wavelengths can also be expected to appear. The trend is toward shorter wavelengths. Low power, room temperature laser diodes that emit continuous-wave radiation at 631 nm and 633 nm have been announced. However, output power will have to increase before these diode lasers can compete directly with helium-neon lasers for applications where a 633 nm wavelength is required. Higher power versions of infrared diode lasers that operate at 900 nm, 1300 nm and 1500 nm may also appear, given the increasing interest in infrared wavelengths for therapeutic use. Diode-laser-based wavelengths will increasingly appear via frequency doubling of diode lasers, and by incorporating diode lasers into microchip lasers and other types of diode-pumped solid-state lasers such as Q-switched Nd:Yag lasers.

4   Laser beam delivery systems

Free beam, hot tip and other contact devices

Much of the rapid expansion into clinical application of medical laser systems is due to increased availability of unique delivery devices which allow the laser light to be delivered to the desired tissue target. Delivery devices can create a significant overlap in the effects and applications of the various types of laser systems, remembering that the ultimate mechanism in many applications is tissue heating.

Laser delivery devices have developed into two broad classes of system:

1. Free or air beam systems (non-contact)
2. Contact and other hot tip systems

This is a somewhat arbitrary classification, but generally free beam refers to techniques in which heat is produced through absorption of light by tissue, and contact refers to those devices that generate significant heat in and of themselves with the input of laser energy. The direct heat from mechanical contact of the tip to tissue then causes conduction or contact tissue heating. There is obviously some overlap in classification.

Free beam devices are those of the CO_2 laser, which requires the beam to transmit through space (even though short distances) before being absorbed by tissue. These include applications through laparoscopes, handpieces and microscopes. Fiber delivered lasers use fibers that are able to transmit the light in a divergent or focused beam can also use a free beam technique.

Fibers which terminate in some device such as a metal tip, sapphire probe, ceramic material or even an altered shape of the fiber tip generate a significant amount of heat at the distal end, and are referred to as contact or hot tip devices. They act as a very intense, but precise, thermal knife avoiding all the complications of electrosurgery. Some of these contact devices, such as rounded or chisel sapphire probes, do actually focus some of the laser light so that combination effects may occur. A combination, rounded tip fiber is also available which may be backed off tissue to vaporize or coagulate as free beam, or touched to tissue to cut as a hot tip.

Lasers may be used with either free beam, hot tip or combined techniques, depending on the fiber or device connected to the laser, and relate more to the desired clinical effect on tissue.

Carbon dioxide lasers

An articulated arm is usually required to deliver the beam to the treatment site. The arm is a hollow tube that has several joints, or articulations, to allow it to be somewhat maneuverable and flexible. The articulations contain reflective mirrors that bounce the beam out of the end regardless of its position. Care must be taken with these arms to avoid jarring impacts which can cause misalignment of the mirrors. Both fixed and adjustable mirror arms can be knocked out of alignment.

Fibers have been developed by several companies for CO_2 lasers. Conventional quartz fibers do not work. Though companies should be commended for their development efforts, CO_2 laser fibers are not yet practical for all day to day surgical cases. Even when fibers are developed, they result in a loss of the good qualities of a TEM_{00} beam, such as the mode and power. Fibers, when sufficiently well developed, will be used as attachments to articulated arms in those situations when endoscopic delivery of CO_2 laser is desired.

Several companies have marketed a hollow waveguide, similar to a fiber, for CO_2 laser delivery. This is a slender, hollow tube that propagates the beam via hundreds of glancing internal reflections, until it emerges from the distal end. These waveguides, called 'airfibers' or other marketing names, are about 1.5 mm in diameter and semi rigid. An inner core of ceramic like material is contained in the sheath. If one flexes or bends the fiber the inner core cracks and the fiber is destroyed. The power density is

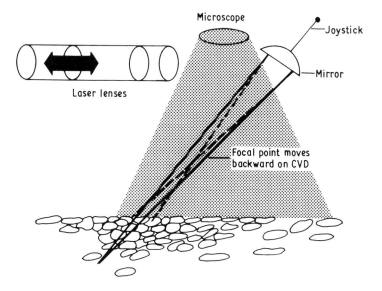

Fig. 4.1   Micromanipulator.

highest within about 1 mm of the tip of the waveguide, and the spot rapidly increases after that.

Various devices can be attached to the end of the articulated arm. A micromanipulator is used to couple the laser to an operating microscope. A lens system in this manipulator allows the spot size to be continuously varied (Fig. 4.1). A joystick is attached to a reflecting mirror which allows the surgeon to direct the beam.

Handpieces allow the freehand use of the laser in the manner of a laser scalpel, though the weight and resistance of the articulated arm does make it somewhat awkward. To defocus the beam the handpiece is simply pulled away from the target tissue (Fig. 4.2). Some handpieces allow the spot to be variably defocused so that the tip of the handpiece is left touching the tissue, but spot sizes enlarged.

Rigid CO_2 laser bronchoscopes and laparoscopes allow use of the laser down the trachea or into the abdominal cavity respectively. These require special scopes and couplers to direct the laser beam down a straight channel.

Computer-laser scanners are used like an electronic micromanipulator on the microscope. This allows the surgeon to outline

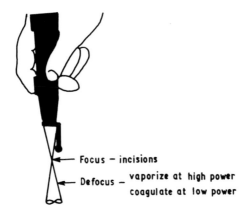

Focus – incisions

Defocus – vaporize at high power
coagulate at low power

Fig. 4.2   Use of handpiece.

any irregular area or shape and automatically lase that pattern. This does serve as a convenience item in some situations, but does not expand laser use over that which can be done manually.

Argon frequency doubled Yag (KTP or other 532 nm) lasers

These are delivered primarily through fiberoptics which may be used as the bare fiber, or terminate into micromanipulators or handpieces with or without contact or sculptured fibers. Fiber tips are cooled by either flowing gas around the tip of the fiber (the fiber is enclosed in a sheath for that purpose), or a bare fiber is used within a fluid environment which keeps the tip cool. Fibers are generally 0.6 mm or 0.8 mm in diameter, but are available in smaller sizes. Adding the sheath increases diameters to roughly 1–2.0 mm.

Fibers are generally used in a non-contact manner. Spot size continuously increases as the fiber is pulled away from the target, causing power density to fall rapidly. Fibers are usually fired about 1–2 cm from the target, but this distance can be varied by varying the power output.

Fibers, for most lasers, can also be drawn across tissue, in direct or near contact with it. The smallest spots and highest power densities are right at the tip, so this allows the fiber to cut and dissect through tissue. Working in a fluid medium keeps the tip cool. This does somewhat abuse the fiber, causing it to burn out at the tip after a short while. Fibers can then be trimmed and repolished at the tip with minimal effort. These fibers may be delivered through standard flexible or rigid endoscopes and laparoscopes. Smaller fibers result in faster cutting at the same powers.

Ophthalmic use of the argon (or krypton, or Q-switched Nd:Yag) is almost always through the slit lamp, similar to a micromanipulator on an operating microscope. Some intraoperative argon ophthalmic procedures use slender probes on the end of fibers for intraocular delivery. Computer controlled scanners are available for the argon, as well as the other lasers used for dermatologic use, such as KTP, dye and copper vapor lasers.

Dye lasers

The 504 nm (green) and 577 nm (yellow) dye lasers are used for their thermal and/or shock wave effects. They are also used in conjunction with small fibers in the same way as the argon and KTP laser. The fibers used for lithotripsy (504 nm) are generally much finer than the standard 0.6 mm argon laser fibers.

The fibers available for photodynamic therapy at 630 nm are used to diffuse the red light uniformly into tissue, rather than concentrating it for its thermal effects. The key aspect of the fiber is the type of diffuser used at the tip. Cylinder diffusers shine light uniformly through the sides of a slender cylinder around the tip. This is used in tubular structures such as the trachea. Light bulb diffusers uniformly spread the light in all directions as with a light bulb and are used in structures such as the bladder. Diffusing lenses shine the light evenly onto the skin. Though not used for thermal effects, the interface of the fibers and diffusers do accumulate some heat, causing the tips to burn up and lose function after a while. A computer scanner is available for more consistent and controlled dermatologic use.

Copper vapor lasers

Delivery systems for the copper vapor lasers are identical to those of the dye lasers for dermatologic work, the primary difference being the availability of different spot sizes from handpieces.

Excimer lasers

In early stages of medical applications of lasers, most of the wavelengths were delivered through conventional fibers. The 193 nm argon fluoride, however, does not pass through regular fibers and currently uses articulated arms. Excimer lasers, used in corneal reshaping work, are computer controlled through an output scanning mirror.

Nd:Yag lasers

These are also fiberoptically delivered lasers, with the exception of the ophthalmic Q-switched Nd:Yag. Standard 0.4, 0.6 or 0.8 mm quartz or glass fibers are either air or liquid cooled. The sheathed fibers have a total diameter of roughly 1.0–2.2 mm. Because of the higher powers of the Nd:Yag laser compared to the argon or KTP, they have acquired a much wider range of clinical applications.

Among the most significant developments in laser delivery systems over the last few years are the ceramic contact probes which include sapphire, used primarily with the Nd:Yag laser, though they could physically function with the argon or KTP laser systems.

Sapphire probes, used only at low powers of around 1–20 W, limit the spread of thermal damage normally associated with the Nd:Yag, and create similar cutting and vaporizing effects as the CO_2 laser. The Nd:Yag laser used with contact probes significantly expands the versatility of this laser.

Contact probes work only when in direct contact with tissue, unlike conventional fiber use. This gives tactile stimulation back to the surgeon and allows for a much shorter learning curve. It also means that underlying tissue will not be affected (Fig. 4.3).

These probes appear to work primarily by the direct heating of the crystal sapphire itself. Regardless of the exact mechanisms of action, they allow for smooth, clean, dry cuts with lateral tissue damage of only about 0.5 mm. Various shapes of probes create different effects such as cutting, chiseling, vaporizing or contact coagulation of tissue (Fig. 4.4). Some of these probes such as the rounded probe do focus the laser energy to produce a combination

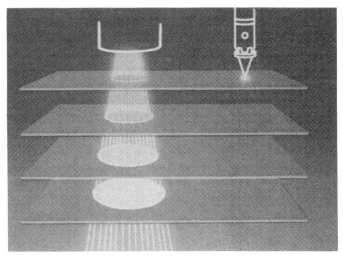

Fig. 4.3   Divergence of the laser from the non-contact bare fiber (left) compared with the non-divergent contact sapphire probe (right), which produces tissue effects only where it touches.

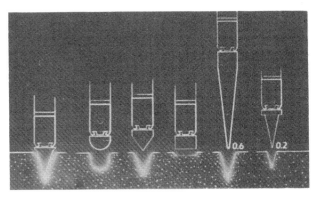

Fig. 4.4   Geometrical shapes of the contact probes and their tissue effects at equal power and time.

effect of non-contact and contact with laser energy transmission into tissue.

Probes must be held in contact with tissue when being used otherwise they will accumulate energy and overheat with potential damage by melting. At higher powers fibers for the probes need cooling either by gas or fluid (both through a sheath), and the probe mount is kept free of tissue accumulation so that the coolant may flow. More newly developed contact probes use lower powers (6–8 watts) and do not require coaxial cooling. Fibers with contact probes have a total diameter of around 1.8 mm or 2.5 mm. They may be used as handheld laser scalpels, or delivered through flexible or rigid endoscopes or percutaneously.

The distal end of 0.6–1 mm bare quartz fibers are sculpted into sharpened fibers for cutting or ball-tipped for coagulation and vaporization to be used in a contact or near-contact mode.

Micromanipulators are available for some Nd:Yag lasers, but their usefulness is limited. When used in this non-contact fashion, the Nd:Yag usually causes diffuse coagulation, in contradiction to the objectives of most microscopic procedures. There are exceptions, and the Nd:Yag can be used with a fiberoptic hand applicator while viewing the field through the microscope.

Aiming beams

Some lasers are invisible to the eye, such as the CO_2 and Nd:Yag, and use a low power, red helium neon or diode laser as the guide light. Nd:Yag lasers also have other colors available as guide lights, and some use a white xenon (non-laser) light. Certain manufacturers of both CO_2 and Nd:Yag lasers have begun incorporating the new multicolor helium neon lasers, and as such can provide red, yellow, blue or a combination of these colors as a guiding light, the idea being that red is hard to see in a bloody field, so that other colors help the surgeon see the beam.

Visible lasers such as the krypton, argon or KTP usually use the laser on a very low setting as its own light guide. With fiber delivery or contact probes, being so close to the target tissue a light guide may be unnecessary.

5

Overview of clinical applications

The surgical and medical applications of lasers are numerous and ever expanding. This introductory guide will not discuss all applications but will attempt to cover the more established procedures where lasers are currently used as a very precise, potentially bloodless 'light scalpel'.

Advantages of laser surgery vary with the surgical procedure, specifications of the laser and the pathology being treated. Realizing the full advantage of laser surgery assumes appropriate application.

Potential advantages of various lasers include:

- Dry surgical field
- Reduced blood loss
- Reduced edema
- Limited fibrosis and stenosis
- Fiberoptic delivery capabilities
- No interference with monitoring equipment
- Potential reduction in spread of metastases
- Precision
- Fewer instruments in the operative field
- Reduced postoperative pain
- Sterilization of the treatment site
- Contact or no-touch techniques

There are many types of lasers, some in routine clinical use such as the carbon dioxide, argon and Nd:Yag, and a whole array of newer wavelengths currently being used for more specific applications in various procedures and surgical specialties.

The CO_2 laser is a commonly used laser in an operating room or outpatient setting. It has widespread applications that make use of its cutting and vaporizing abilities. Its secondary cautery effects are also helpful.

The argon laser is the most prevalent overall, though not in an operating room setting. Its primary use has been in ophthalmology as a retinal photocoagulator. It is also widely used for colored skin lesions in dermatology, though the copper vapor and pulsed dye lasers may be more specific for cosmetic lesions. It is beginning to see greater endoscopic use. The KTP, or other 532 nm lasers, are very similar in tissue effects to the argon laser. However, there are subtle differences in some areas of application.

The Nd:Yag laser, near infrared at 1064 nm, is widely used in surgery. Its fiberoptic delivery, high power when needed, and with contact probes or sculpted fibers for fine cutting and vaporizing, makes this a very versatile instrument in many specialties. General surgery, which previously saw little laser use otherwise, has rapidly expanded in both open and laparoscopic surgery. The use of Nd:Yag in ophthalmology as a Q-switched device has seen explosive growth in previous years and is now routinely utilized often in an outpatient or doctor's office setting.

Mid-infrared lasers such as the Erbium:Yag at 2.94 μm may see increased applications in orthopedics and dentistry as a highly precise bone saw and drill. Other mid-infrared lasers such as the Holmium:Yag have tissue effects that combine some of the vaporization effects of the CO_2 laser with the ability to be transmitted through standard quartz fibers, leading to its current use in arthroscopy.

Though laser surgery can result in better patient care – with the type of laser usually specific to the task – many of the practical advantages of using various lasers for surgical procedures are not always due to better results or patient comfort. Many good applications are for the convenience and preference of the surgeon. Having one instrument (the laser) which accomplishes the same tasks as several mechanical instruments can save time, reduce instrument swapping, and in general make the procedure less hassle and easier to perform. Because of this convenience aspect of lasers, one will find that several types of lasers may be utilized for the same procedure, based on the preferences of the surgeon. Some people like to drive cars with stick shifts, and others prefer automatic transmissions.

Lasers are another surgical tool for a physician to carry out more precise surgery be it incision, excision, coagulation, vaporization or combinations of these effects on normal and pathological tissue. Lasers can replace the cold steel scalpel or electrosurgery. More importantly lasers can be linked to other diagnostic technologies such as endoscopy (flexible or rigid), ultrasound, CT scanning, magnetic resonance imaging (MRI) or positron emission tomography (PET) scanning. This converts a diagnostic procedure into a therapeutic procedure and allows the whole field of minimal (MIS) or less invasive surgery (LIS) to expand rapidly into patient and healthcare benefits. The following is an overview of major specialties that currently use lasers.

Gynecology

This specialty is one of the major users of lasers in a wide variety of gynecological conditions ranging from external genital procedures, to laparoscopy and hysteroscopy.

The carbon dioxide, argon, frequency doubled Yag (KTP) and Nd:Yag are used depending on the surgeons' preference, instrument availability and pathology being treated. Each system offers certain advantages in its ability to cut, vaporize or coagulate tissue.

Laser laparoscopy yields a significant advantage in providing a method to treat mild to moderate endometriosis at the time of the diagnosis. It can vaporize or coagulate endometriomas and dissect adhesions. It is used for restoring patency to obstructed fallopian tubes and treating ectopic pregnancies. The carbon dioxide laser has been the primary laser used for laparoscopy. Older articulated arms and laser couplers were very awkward, but newer units eliminate many of the previous problems. Excessive smoke production requires a continuous high flow insufflator to clear the smoke and maintain a pneumoperitoneum. Fiberoptic laser systems are the trend in laser laparoscopy. They are easier to work with, eliminate much of the smoke and do not require special instrumentation. These laser systems include the argon, KTP and Nd:Yag with the contact probes.

Abdominal cysts and tumors may be excised or vaporized with similar benefits. The ability to vaporize tissue with minimal surrounding damage, even when associated with dense and extensive tissue adhesions, has proven of great value for patients with complications of pelvic inflammatory disease (PID). The laser is used in microtuboplasty to cut the tube prior to reanastomosis. In contrast to electrosurgery, which is very suitable in many circumstances, laser provides a higher degree of control and more precise ablation or excision.

The benefits of this are more controversial but it does eliminate bleeding. A laser incision also provides a precise, atraumatic means of opening the end of a closed tube in a bloodless fashion. In neosalpingotomy, the power density is lowered and the laser used to 'paint' a ring around the end of the tube. The shrinkage causes a flowering of the tube. This is much more expedient than conventional suture technique, but there is still some question as to the longevity of results with the laser technique. When extensive adhesions are encountered, particularly in the cul-de-sac on the bowel, laser can reduce operative time.

Uterine myomas may be removed by vaporization or excision. A microlaser myomectomy provides hemostasis and precision when removing fibroids.

Laser has been used in cornual reimplantation. Radical procedures, such as radical vulvectomies, and excision of large vascular tumors are performed with laser for better hemostasis. These applications require higher power from a CO_2 laser (50–60 W), or contact techniques with a Nd:Yag laser.

One of the most common uses of the CO_2 laser is the treatment of cervical intraepithelial neoplasia (CIN) for either ablation or excision. This is most commonly done through the colposcope. The Nd:Yag laser with contact techniques will perform an adequate excision or ablation.

Laser vaporization of the cervix provides advantages over cauterization, knife conization or cryotherapy. It leaves the cervix in a more viable condition, with no stenosis and minimal scarring. It is possible to tailor the treatment to the extent of the disease and give assurance that the entire diseased area was treated. It eliminates the heavy discharge associated with cryosurgery and is significantly more precise. Laser colposcopy is an outpatient or office procedure for ablations – a significant advantage.

Similar advantages are gained in the laser treatment of vaginal intraepithelial neoplasia (VIN).

The use of the 'LEEP' electrosurgical procedure could replace much of the currently performed CO_2 laser work on the cervix because of the very low cost and ease of performance of the electrical procedure. However, the CO_2 laser procedure remains more precise when done correctly, and is still the choice for any compromised cervix, and vaginal and vulvar work.

Genital warts such as condyloma accuminata may be treated to advantage with the laser. The CO_2 laser not only vaporizes the lesions (frequently through the colposcope) but can also flash sterilize the skin between lesions. This kills the latent virus and reduces frequency and extent of outbreaks. Argon or KTP lasers may be used for superficial lesions through the colposcope. Contact probes on the Nd:Yag have been used to superficially vaporize the warts in a contact fashion.

In the future, HPV may be treated in a non-surgical fashion with photosensitizing drugs and a laser to provide the treatment light. (See photodynamic therapy in Chapter 5).

Hysteroscopic laser procedures are increasing in popularity to avoid the more standard hysterectomy. Dysfunctional uterine bleeding (chronic menorrhagia) can be treated using the Nd:Yag laser to coagulate the endometrial lining on an outpatient basis. Not all women become totally amenorrheic after treatment but most are satisfied with the result. Other intrauterine surgery using the contact Nd:Yag, KTP or argon lasers include division of uterine septa and synechia, and the excision of intrauterine fibroids. Laparoscopic assisted hysterectomies with the laser are now being performed to decrease the morbidity of a conventional abdominal or vaginal hysterectomy.

Otorhinolaryngology

This is one of the best uses for the laser because of the no-touch technique, long 'reach' of the laser, absence of postoperative swelling or stenosis, dry operative field, and greatly reduced postoperative pain. This is true of the CO_2, argon and 532 nm lasers when coupled to the micromanipulator.

Applications of lasers include microlaryngeal surgery, treatment of diseases of mouth and tongue, otology, nasal and sinus surgery. Laser sinus endoscopy has the advantages of being either contact or non-contact, with absence of postoperative swelling or stenosis, a dry operative field and reduced postoperative pain.

Application of the CO_2 laser to laryngeal diseases requiring microlaryngoscopy has provided a degree of precision otherwise impossible. The argon or 532 nm lasers may also be used through a micromanipulator to achieve this same no-touch precision. The precision cutting and excellent healing is an even greater advantage than the hemostasis. Postoperative pain is minimal and most procedures may be done on a same day surgery basis. The need for tracheotomy is reduced. These cases require general anesthesia and appropriate safety precautions must be taken because of the flammability of the endotracheal tube.

Surgical applications of the laser in microlaryngoscopy include vocal cord nodules, polyps, hyperkeratosis, granulomas, arytenoidectomy, Reincke's edema, cysts, webs and laryngeal stenosis.

All airway lesions are more critical in the child than the adult because of the size of the airway. The treatment of congenital and acquired lesions in pediatric surgery has proven the laser to be a remarkably effective tool. It is ideal for infants. Its properties of hemostasis, enhanced visibility, lack of postoperative edema and scarring all contribute to its successful application in pediatrics.

Recurrent respiratory papillomatosis occurs throughout the anterior nasal cavity, subglottis and mainstem bronchi. These relatively inaccessible locations make the laser ideal for removal of all visible papillomas by vaporization. Complete hemostasis allows all visible lesions to be destroyed under constant visual control. There is minimal damage to underlying tissue and the airway can be maintained so that tracheotomy is usually unnecessary. Recurrences are not eliminated, but a large percentage of patients go into a year or more of remission after two or more excisions with the laser.

Lasers may be used in intranasal procedures such as turbinectomy, choanal atresia, and telangiectasias. The contact Nd:Yag laser works better than the carbon dioxide, KTP or argon lasers for most intranasal cutting and vaporizing because of the vascularity of the tissue involved. Lasers effectively treat polyposis, synechia and granulomas. For those patients with rhinophyma, the contact Nd:Yag or carbon dioxide lasers have been successfully used to treat the condition, with preservation of dermal elements for reepithelialization of the nasal surface.

The use of lasers in sinusoscopy has been very useful. The slender fibers may be easily manipulated through a flexible or rigid scope, and passed into a tiny meatus. The argon, 532 nm lasers, contact Yag and Ho:Yag lasers all work well for sinus work. The Ho:Yag laser has been shown the best for actual bone drilling. Laser tonsillectomy is most appropriately carried out in patients with coagulopathies such as hemophilia. Proponents of laser tonsillectomy on normal patients point out improved hemostasis and reduced postoperative pain. The fiberoptic systems, used in contact with the tissue, are the most convenient for this application, though CO_2 lasers may also be used. Patients have less postoperative pain and swelling, and are able to begin eating earlier.

Lesions of the oral cavity, such as leukoplakia and other benign lesions, may be excised or vaporized and the frenulum incised for tongue release procedures. Partial hemiglossectomies and radical neck dissections are carried out with various lasers for carcinomas in the head and neck regions. Thyroid surgery be it nodule excision, lobectomy or total thyroidectomy for carcinoma have been evaluated with good results relating to hemostasis, laryngeal nerve and parathyroid gland preservation. The use of lasers in otology is still being debated regarding whether the carbon dioxide or green light lasers are better for such procedures as stapedotomies and tympanoplasty revisions. The small spot size of the laser appears to give greater precision in the treatment of ear conditions.

Pulmonary medicine

The laser is precise, somewhat hemostatic and immediately vaporizes the obstruction. A rigid CO_2 laser bronchoscope, and coupler cube, is required for delivery of the beam. The long focal length of the laser lens also causes the beam to remain in focus for a long distance. This is a potential hazard and one must carefully avoid penetrating the trachea and underlying great vessels by using moderate power levels and control of exposure time by pulsing (timed or gated pulses) techniques.

The Nd:Yag is used more effectively to treat airway obstructions than the carbon dioxide laser. The Nd:Yag can be used with the bare fiber to achieve coagulation and the subsequent debridement of necrotic tissue may be accomplished with instruments through a rigid bronchoscope. Contact probes may be used to precisely create a lumen through a tumor. A combination of air and contact laser energy delivery may be necessary.

A flexible bronchoscope may be used with the Nd:Yag fiber for access to bronchi beyond the carina. However, when the tumor is more accessible, a rigid scope is preferred because of the greater ability to debride tissue, better suction and manipulation at the tip of the scope. A special rigid bronchoscope is available from several companies that has been modified for Nd:Yag laser use by swiveling the proximal port and providing channels for fibers.

One must be aware of the deep coagulation the non-contact Nd:Yag can achieve. Powers are frequently limited to around 25 W for 0.5 s for safety reasons since most tumors would be overlying the carina or large vessels around the bronchi. The coagulation could extend into the underlying vessel, causing delayed necrosis and perforation within one or two days. A good visualization of the anatomy, avoidance of the periphery of the tumor, and use of lower power helps prevent this. Contact probes limit the damage to about 0.5 mm. Tumors that occupy the mid portion of the trachea may be handled more aggressively. Extraluminal tumors that compress the airway cannot be treated with any laser technique by bronchoscopy. Lasers are also used for treatment of vascular lesions, release of strictures, opening of stenoses and photodynamic therapy (PDT) for both squamous and adenocarcinoma of the lung.

Neurosurgery

The carbon dioxide laser is an ideal instrument for microscopic use, due to its precision and long reach into small holes. The Nd:Yag is a very useful adjunct in shrinking vascular tumors, and for treating aneurysms and arteriovenous malformations (AVMs). Contact probes on handpieces, used while viewing through the microscope, offer a more precise way to use the Nd:Yag for creating hemostasis in small feeder vessels. Most craniotomy openings may be made smaller when a laser is used. An orange-sized tumor could be removed through a quarter-size opening by coring through the center of the mass. Fewer instruments are in the field of view and this, coupled with the good hemostasis, makes it much easier to identify the anatomy. Laser-treated patients appear to be more alert postoperatively than with conventional techniques.

Meningiomas which have tough dural attachments may be easily 'peeled away' with a laser. The use of lasers for acoustic neuromas has become a standard due to their precision and preservation of the acoustic nerve. Lasers are similarly ideal to preserve function when treating tumors around the optic chiasm and nerve. Tumor remnants may be shaved off one layer at a time from an artery, with no damage to the underlying vessel. The Nd:Yag laser, when used in a non-contact manner to desiccate and coagulate large tumors, causes the tumor to shrink and peel away from normal brain tissue.

In transsphenoidal hypophysectomy, the laser offers an atraumatic, no-touch technique that eliminates instruments in the narrow canal of the speculum. Recurrent pituitary adenomas may be more easily removed with the laser. New areas include use of flexible and rigid endoscopes (encephaloscopes) for laser energy allowing less invasive surgery through smaller delivery incisions. Lasers are also being used with real time MRI to treat inoperable and centrally located tumors. Combined CO_2–Nd:Yag lasers with computer guided systems for special control of the laser beam are being investigated using CT, MRI and ultrasound imaging.

Spinal tumors also benefit from laser surgery, particularly intramedullary tumors. Manipulation of the cord is kept to a minimum, resulting in less damage to both cord and nerve roots. Lasers may be used for fenestration of syringomyelia to offer a permanent fluid pathway. Intractable pain has been treated with laser lesions of the dorsal root entry zone (DREZ) in a more precise and controlled method than other techniques. The laser can be used to dissect away muscle, avoiding the muscle spasm seen with electrocautery and reducing pain following spinal surgery.

Back pain and neurological defects due to prolapsed intervertebral discs in the lumbar region are being treated on an outpatient basis using a percutaneously placed fiberoptic delivery system with either the Nd:Yag or KTP or newer holmium lasers. This percutaneous laser disc decompression (PLDD) is successful in over 80% of carefully selected patients and will have a significant effect on healthcare costs in the future.

Dermatology

Since the 1960s lasers have been used extensively in dermatology and most wavelengths have found applicability depending on the underlying pathological process. The CO_2 and argon lasers are used extensively in dermatology. KTP and other 532 nm lasers have similar applications as the argon. The Nd:Yag may be good for cavernous types of hemangiomas and keloid revisions. Tunable dye lasers, either continuous wave or pulsed, are the current trend for very selective use in cosmetic type vascular skin lesions. There are advantages to each type of system depending on the type of lesion treated.

The pulsed dye laser at 577–585 nm (yellow) is used as a very selective photocoagulator. The wavelength, coupled with short pulse widths, achieves precise vascular coagulation beyond the abilities of the continuous wave argon. The copper vapor laser, producing yellow light at 578 nm or green light at 514 nm, does the same thing but with a different pulsing configuration. These lasers, because of color selectivity, are used to photocoagulate pigmented cutaneous lesions such as portwine stains, capillary hemangiomas, telangiectasia, strawberry marks, Campbell DeMorgan senile angiomas, and acne rosacea. It may be used to remove tattoos, treat pyogenic granuloma, sebaceous nevi and the Peutz–Jegher syndrome. Less established uses include keloid scars, subcutaneous varicose veins, road skid burns, moles, warts and nevi of the Osler–Weber–Rendu syndrome.

The KTP laser offers a high frequency, higher peak power, greener beam than the argon. Its pulsing characteristics can create a 'cooler' photocoagulation of small vessels in sensitive areas of skin and can be used with a computer-driven scanner.

Portwine hemangiomas involve an increase in the number of vessels in the subepidermal zone. Lasers, except for the CO_2, transmit through the epidermis as if through a window pane, and coagulate vessels within the dermis. The treatment is performed as a series of applications over several months, resulting in gradual fading. Test areas are first performed to determine which laser, and parameters, provide the best cosmetic result. The pulsed yellow light lasers, pulsed dye and copper vapor are probably the best systems for portwines. They have decreased pain and scarring to such a degree that it is reasonable to treat small children, and even babies, with good results and little discomfort.

Tattoos may be removed with a laser to give good, but not perfect results. Visible lasers selectively obliterate the dye in a tattoo. Green argon or KTP is used for black and red dyes, and red krypton works with blue or green dyes. Professional tattoos are easier to remove than amateur ones because of the consistency in dye depth. The visible lasers shine through the epidermis, leaving no surface scar. Hypertrophic scarring may be a problem and though the tattoo can be erased, a scar may be left in its image.

The CO_2 laser has been used in removing tattoos. It provides a uniform vaporization without regard to dye color or shape. Although a superficial scar is left, it can be blended at the edges so that no tattoo image remains. It is important to not relase through char, which could cause significant scarring.

Recently tattoos have been successfully removed with the Nd:Yag laser. Low levels of Nd:Yag light have been shown to decrease collagen production in fibroblast cultures. This has implications for both reduced scarring, and in the possible excision and treatment of keloids. Keloids have been successfully treated that were refractory to steroid injection and excision.

The original laser developed in 1960 was the ruby laser, producing a pulsed output of around 694 nm deep red light. This laser is now seeing a resurgence for the treatment of tattoos and removal of age spots. The high peak power pulses can obliterate the pigment underneath the surface without burning over- or underlying tissue. However, it is not very effective against red pigment. Newer solid state lasers include Q-switched Yag with frequency doubling which may provide therapy for vascular, pigmented, tattoos and viral skin lesions.

Cosmetic, plastic and reconstructive surgery

The skin lesions referred to under the dermatological section are often treated by plastic surgeons as well. The ability of the laser to vaporize, incise and remove tissue with minimal bleeding, reduced edema and tissue damage are of considerable importance in cosmetic and reconstructive surgery.

Breast surgery including reduction mammoplasty, augmentation or cancer procedures can use the laser as the surgical tool. The laser seals lymphatics and may be of benefit in treatment of cutaneous malignancies. The carbon dioxide laser has been used in breast surgery to reduce blood loss. Skin incisions made with the carbon dioxide laser produce scars that are cosmetically similar to that of a cold knife but the articulating arm and slowness have delayed its use. The contact Nd:Yag scalpels are excellent for use in breast surgery. The skin incision may be made with a knife and the rest of the dissection performed with the laser. Hemostasis is excellent, with minimal thermal damage to surrounding tissue, less bleeding, reduced pain and postoperative drainage.

Contact probes or sculpted fibers with the contact Nd:Yag are used for raising skin flaps. The dissections are very clean and dry with no charring. Carbon dioxide lasers may also be used, but it is more difficult to control depth of penetration with the CO_2 handpiece, resulting in damage to the flap. Reconstructive surgery including vascular malformations and cleft palates are giving improved results with lasers. Cosmetic surgery with lasers includes blepharoplasty and face lifts. Laser assisted liposuction is being evaluated to decrease the intraoperative blood loss and postoperative complications.

Gastroenterology

The argon, KTP and Nd:Yag lasers have been used endoscopically in the treatment of gastrointestinal disease with the Nd:Yag being the primary treatment modality.

The laser may be used in the endoscopic treatment of bleeding from peptic ulcers. Hemostasis may be achieved by using the laser fiber either in a contact (with probes or sculpted fibers) or non-contact fashion. In cases which are actively bleeding at the time of endoscopy, and are situated in an area that is difficult to access for contact or heater probes, the laser offers the advantage of being able to coagulate without touching. A jet of coaxial CO_2 gas is delivered down the fiber sheath to cool the tip and clear blood from the field of view. Contact Nd:Yag lasers used with fiberoptic endoscopy avoid overdistension of the gastrointestinal tract, use much less power and offer the advantage of mechanical pressure to coapt blood vessel walls.

Bleeding from arteriovenous malformations, gastric erosions, bleeding peptic ulcers, Osler–Weber related lesions and the watermelon stomach have all been successfully treated, often reducing blood transfusion requirements and avoiding major resectional surgery.

Recanalization of advanced, obstructing tumors in the esophagus, antrum of stomach, colon and rectum is an excellent use of the Nd:Yag laser as a palliative therapy. It can provide relief of symptoms, particularly for the dysphagia associated with advanced esophageal and other tumors, unsuitable for conventional forms of treatment. Obstructions may be coagulated in a non-contact fashion, then debrided, but care must be taken to avoid perforation. Photodynamic therapy (PDT) is being evaluated in the palliative treatment of carcinoma of the esophagus.

Obstructions may also be opened with the use of contact vaporizing probes, with coaxial water. Limited lateral thermal necrosis decreases the concern for fistulas. Sessile polyps and villous adenomas have also been successfully treated with the laser, thus avoiding major colonic resections.

Urology

The Nd:Yag laser is the primary instrument for endoscopic procedures, and the CO_2 laser for external lesions. Argon or KTP may also be used endoscopically.

The primary use of the Nd:Yag laser in urology has been the transurethral treatment of multiple, superficial bladder tumors and occasionally muscle invasive bladder tumors. The laser fiber is passed through the cystoscope with a fiber deflector on the bridge with the bladder distended. The laser causes necrosis of the bladder tumor down to the serosa without seriously compromising the mechanical stability of the bladder wall or causing perforation. The photocoagulation may be performed in either a non-contact mode or used with the contact probes. Transurethral catheter drainage of the bladder is eliminated and the procedure itself does not take long to perform. Tumors up to 2 cm may be totally eliminated by the laser while larger ones may be removed via a cutting loop and the tumor bed then coagulated by the laser. Second sessions for larger tumors may be necessary. The laser is especially useful for small, multifocal bladder tumors.

Stone lithotripsy is performed using pulsed dye lasers. A transurethral approach is necessary and a very slender fiber is advanced until it abuts up to the stone using a miniscope for visualization. The laser impact causes a shock wave that mechanically disintegrates the stone. Other wavelengths being developed include alexandrite, titanium sapphire and Q-switched Nd:Yag lasers.

The carbon dioxide laser can vaporize and sterilize condylomata and other external genital lesions.

The Nd:Yag laser has also been used clinically to treat penile carcinoma and prostate cancer. Prostatectomy or prostatotomy with the laser is being evaluated and several options are available, from limited resection, balloon hyperthermia, to lateral firing laser fibers, to photocoagulate the prostatic bed. A lateral firing fiber with the Yag laser is now developed for treating benign prostatic hypertrophy avoiding resection, bleeding and prolonged catheterization. Urethral and bladder neck strictures have been successfully treated with both the argon and Nd:Yag lasers. The contact probe with the Nd:Yag laser is an excellent method for treatment of strictures.

Partial nephrectomies may be performed with the contact Nd:Yag laser to reduce blood loss and retain function in the remaining portion of the kidney. One of the experimental uses for low power density lasers is for tissue welding, e.g. reanastomosis of the vas deferens. PDT is being evaluated for the treatment of multifocal superficial bladder cancer. Early results reported bladder contraction due to fibrosis. Newer laparoscopic procedures include pelvic lymph node dissection for staging of prostate and bladder cancer and nephrectomies.

Maxillofacial and dental surgery

Laser work in these fields is primarily divided into soft tissue such as the gum and hard tissue, e.g. enamel and dentin. The 1990s is seeing an explosive growth of the use of lasers in the field of general dentistry and periodontics.

The introduction of low average power, pulsed Nd:Yag laser, delivered through tiny fibers, has allowed the development of many periodontic applications. This laser is being used to sterilize gum pockets, with significant long-term reduction in bacteria and reduction in pocket sizes. It is similarly used to sterilize root canals – the fiber size is only a fraction of a millimeter. It is claimed to firm up gums and strengthen root attachments. Its use in bloodlessly incising and debulking soft tissue, such as gum, overlaps to some extent with the CO_2 laser; it is very effective but a bit slow. The need for local anesthesia and postoperative analgesics has been significantly reduced, or even eliminated in some patients. It is being successfully investigated clinically to vaporize caries and prepare the bed for amalgams – again with patients requiring no local anesthetic. This is especially useful in treating caries in children. Any significant debulking still requires a local block because of the time required for the procedure which allows for heat to build. The psychological impact to the patient of eliminating the whine and vibrations from a drill will be a significant factor in its widespread acceptance. In most instances, this laser creates less heat on the tissue than the drill itself. Some unique applications of this pulsed laser are also being examined, such as its ability to 'seal' the microtubules in temperature sensitive teeth as a form of treatment. It can also etch enamel and dentin.

The CO_2 laser is being used clinically to perform gingivectomy. This laser seems to perform better on soft tissues such as gum and buccal mucosa compared to the Nd:Yag modalities. It has been particularly useful in treating patients with dilantin hyperplasia of the gums because it reduces or eliminates bleeding, sterilizes as it vaporizes, and results in significantly less postoperative pain than conventional techniques. Additionally, the spot may be defocused to vaporize and 'sculpt' areas of gum rather than just incise. One must be careful not to mark underlying teeth with the laser. An osteal retractor placed between gum and teeth may serve as an adequate backstop. The contact probe on the Nd:Yag laser has been shown to incise soft tissue without marking the underlying enamel. The use of the contact Nd:Yag laser is seeing clinical use in dentistry for soft tissue incision and sculpting. Temperomandibular joint (TMJ) surgery is performed with the Ho:Yag, or contact Nd:Yag laser.

The Holmium:Yag, Erbium:Yag and various excimers may be used in dentistry in the future because of their ability to cleanly etch and vaporize hard tissues such as enamel.

General surgery

This is the area of most rapid increase in laser utilization over the last several years with many more procedures being carried out with the laser as the surgical tool, replacing the knife and electro-surgery. Although the carbon dioxide laser was initially used, the contact probes and sculpted fibers with the Nd:Yag have significantly expanded the use of lasers in the traditional areas of incision, excision, vaporization and coagulation. Other laser wavelengths include the KTP and holmium.

The laser has been used in the excision or vaporization of liver metastases, to perform anatomical and non-anatomical resections of the liver, and in performing pancreatectomies and other vascular organ work. The benefits include improved intraoperative hemostasis, less tissue damage and more rapid postoperative recovery with fewer complications.

The laser sterilizes as it vaporizes so it is useful for debridement of external ulcers. It is also used in burn debridement to achieve sterility and hemostasis and the treatment of pilonidal cysts.

Contact Nd:Yag laser surgery is finding increasing applications in breast surgery, gallbladder, hernia repair, thyroidectomy and hemorrhoid operations. Advantages appear to be less bleeding and postoperative pain with earlier mobilization.

Laparoscopic procedures have rapidly advanced, including laparoscopic cholecystectomy, hernia repair, appendectomy, vagotomy, lymph node dissection and biopsy of malignancies. Although the benefits of the laser over electrosurgery include lack of interference with monitoring equipment and pacemakers, and avoidance of burns and 'spark' injuries, the cost of the laser and education are causing some confusion as to which modality is best. Obviously the laparoscopic 'key hole' surgery is the reason for the successful outpatient to one-day stay procedure.

The laser is important to any laparoscopic procedure because of its convenience and precise effects. It can perform the task of several instruments and is a good way to hemostatically dissect and vaporize while retaining a high degree of control over lateral heat damage. Because of the rapid proliferation of laparoscopic cholecystectomies, many general surgeons have discovered that electrosurgery (like a hook tip) is just as effective and cheaper than laser. One extrapolation of this observation is that laser has no real value for them. Our advice is to not throw the baby out with the bathwater. The use of an electrosurgical probe, though not as precise as laser modalities, does work in dissecting the gallbladder away from the liver bed. The extra precision of the laser may offer many advantages, and there are many other areas where electrosurgery is not desirable. As general surgeons expand their laparoscopic procedures however, we think they will find many more situations where the precision, lack of muscle stimulation, predictability and ability to use tiny fibers with a laser will greatly expand their effectiveness and safety in performing laparoscopic procedures. The CO_2, argon, 532 nm lasers, Nd:Yag and Ho:Yag lasers have all been used for laparoscopy.

Hemorrhoids have been treated successfully with both CO_2 and Nd:Yag lasers. The CO_2 laser is used primarily for externals and skin tags, and the Nd:Yag for shriveling feeders to the internal hemorrhoids. Contact probes in the Nd:Yag may be advantageous here. Techniques vary widely. Some methods combine advantages of cryosurgery, laser and banding techniques. Other techniques utilize the contact Nd:Yag laser and provide for a local block so that the procedure is performed in an outpatient clinic.

In various types of mastectomies the contact Yag achieves less intraoperative bleeding, thinner skin flaps and sealing of lymphatics. Axillary clearance of lymph nodes becomes a cleaner anatomical dissection of the axillary vein and related nerves in areas where electrosurgery is not used. Small clinical series of randomized prospective studies indicate less postoperative analgesia, earlier mobilization and decreased length of stay.

Orthopedics

Lasers have finally found a firm place in the armamentarium of the orthopedic surgeon for arthroscopy. The pulsed Holmium:Yag laser is proving to be a very effective tool for shoulder and knee arthroscopies and carpal tunnel work. Contact Nd:Yag, KTP and CO_2 lasers may also be used.

Other lasers have been used previously for arthroscopy with varying degrees of success, but all seemed to be more trouble than they were worth. The fiberoptic delivery of the Ho:Yag through fluid allows the arthroscopy to proceed in the normal fashion. Because of its 2060 nm mid-infrared wavelength, and the high energy pulses, the tissue effect approximates the precise cutting and ablation of the CO_2 laser wavelength, and provides for good hemostasis – all under fluid and through a slender fiber. The small fiber handpiece allows good reach into difficult areas such as the posterior horn of the medial meniscus, without scuffing of articular surfaces in getting it there. In addition to hemostatic incisions, it can sculpt, vaporize and shape tissue like no existing tool. The fluff from chondromalacia may be smoothed and contoured very easily with this laser.

Recent developments with ceramic contact probes attached to the Nd:Yag laser also provide for good cutting effects arthroscopically, but cannot be defocused for sculpting and vaporization.

The CO_2 laser can provide for versatile arthroscopic use, but its disadvantages are its inability to work through fluid, or to be transmitted through flexible fibers. This means that the direct coupling of the beam through the channel of a probe, and working under the encumbrance of CO_2 gas joint insufflation becomes somewhat of a hassle. Recent techniques allow for bubbles to be produced at the end of the laser probe, so that the laser will work through the gas bubbles and still allow for fluid irrigation of the joint. Other lasers such as various excimers are being used for investigative work for arthroscopy, but the Holmium:Yag laser is currently the laser of choice for arthroscopy.

The CO_2 laser has been used both with a handpiece and through the microscope to vaporize polymethylmethacrylate in the shaft of the femur when replacing artificial joints. Irrigation is necessary to avoid flaming, and high smoke evacuation eliminates the noxious fumes. The laser is best used to core the glue, leaving only a thin crust to chisel out. This is a much more atraumatic method of removing glue from deep within the bone but the technique is not well accepted due to the noxious fumes.

Percutaneous laser disc decompression (PLDD) is being performed with the Nd:Yag, KTP and holmium lasers. Several hundred patients have now been successfully treated with the Yag laser fiber technique performed percutaneously under radiological control. Both the short- and long-term follow-up indicate that in selected patients many minor spinal surgeries will be avoided by this outpatient procedure.

An exciting prospect for reconstructive orthopedic microsurgery is the use of laser to achieve tissue fusion (welding). This has been done with low power density CO_2 lasers and the argon lasers. A 1.3 μm Nd:Yag laser, with programmed dosimetry is being investigated for laser fusion on several types of tissue. Best results have so far been shown on vessels and nerves.

Several lasers, including the Erbium Yag (Er:Yag) and excimer are being studied for use in cutting bone. Fiberoptic delivery of the erbium can be accomplished by transmissive sapphire fibers.

Thoracic surgery

The endoscopic applications for the palliative treatment of obstructing and bleeding bronchial carcinoma have been referred to in the section on pulmonary medicine.

Lasers like the CO_2 and especially the contact Nd:Yag are used in performing a thoracotomy and partial pneumonectomy. Reported benefits include less pain and reduced air leakage with greater lung preservation. Insertion and removal of pacemakers is easily accomplished with lasers avoiding the electrical interference obtained with electrocautery.

The newer areas of video-thoracoscopic procedures include vagotomy, thoracoscopic blebectomy by wedge resection or treatment of spontaneous pneumothorax and thorascopic node dissection for management of pleural malignancies. Transthoracoscopic cervical sympathectomies have been performed for hyperhydrosis of upper extremities.

Pediatric surgery

All the indications for laser surgery in adults are applicable in neonatal and pediatric surgical specialties. These include treatment of colored cutaneous lesions. In the past, portwine stains and other vascular lesions could not or would not be treated by a physician until patients were in their teens. With pulsed dye lasers there is less damage to surrounding areas and they work better on lightly pigmented lesions. Papilloma viral lesions in the oral cavity can be treated at an early age with the argon laser.

Abdominal operations, thoracic, urological, otorhinolaryngeal, reconstructive and orthopedic surgical procedures can all use lasers with similar benefits of reduced intra-operative bleeding and decreased tissue damage leading to reduced post-operative morbidity.

Plate I Patient undergoing photoradiation therapy with a dye laser for a tumor in his right cheek. Four fibers are delivering laser light which can be seen shining through the skin.

(Photograph by courtesy of J. McCaughan, Grant Hospital, Columbus, Ohio).

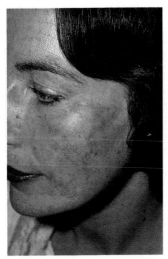

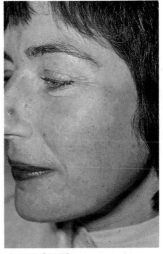

Plate II(a) Patient with a portwine stain (birthmark) on her cheek.

Plate II(b) The same patient after treatment with an argon laser, showing the greatly improved cosmetic appearance which can be achieved.

(Photographs by courtesy of Mr J.A.S. Carruth, Royal South Hants Hospital, Southampton).

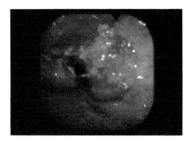

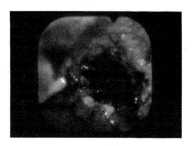

Plate III(a) Tumor of the esophagus. The tumor is the bright red mass in center, and has almost completely blocked the esophagus leaving only a very small opening (the black area to the left of the tumor).

Plate III(b) The tumor has been removed by treatment with a Nd:YAG laser thereby clearing the blockage.

(Photographs by courtesy of Dr S.G. Bown, University College Hospital, London, and first published in *The Proceedings of the Royal Institution of Great Britain*, 55, 177–97, 1983).

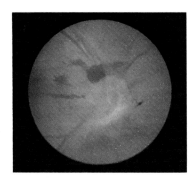

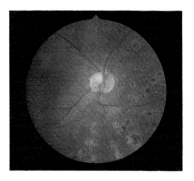

Plate IV(a) Proliferative diabetic retinopathy in an insulin-dependent diabetic patient of 20 years duration. Several areas of hemorrhage with optic disc neovascularization (new vessels) can be seen.

Plate IV(b) Complete regression of the hemorrhage and new vessels has been achieved by pan retinal photocoagulation with an argon laser.

(Photographs by courtesy of Mr R.J. Cooling, Moorfields Eye Hospital, London).

Podiatry

Lasers are used in the treatment of neuromas, plantar warts, soft tissue dissection and bunionectomies. The laser is also useful in patients where it is not advantageous to apply a tourniquet such as patients with circulatory disorders or diabetes.

Ophthalmology

Ophthalmologists were the pioneers of laser surgery. Lasers have been used for precise photocoagulation of the retina since mid-1960. The argon laser is the primary photocoagulator. Krypton lasers, with their yellow and red wavelengths, are also used by retinal specialists to achieve greater control in the macular area. The Q-switch, and to a lesser degree mode-locked, Nd:Yag lasers are also used in ophthalmology. All these lasers may be used as stand-alone units, or commonly the argon and krypton are combined in one unit and delivered through the same slit lamp and mated with the Q-switched Nd:Yag. New, compact diode lasers are finding a niche for photocoagulation purposes when laser size is a consideration.

The use of the carbon dioxide laser has been very minimal. As a vaporizing instrument it has been used to excise scleral tumors and has also been used to create bloodless scleral flaps. The wavelength of the CO_2 laser cannot be transmitted into the eye, in contrast to the argon and Nd:Yag, and therefore it is only rarely used for open procedures.

Argon is the common ophthalmic laser, used in an office or clinic setting, through the slit lamp, or intraoperatively with intra-ocular probes. Patients with diabetic retinopathy suffer from a proliferation of blood vessels on the retina which are fragile, tend to bleed, and may even tear the retina because of retraction of the vitreous membrane. Panretinal photocoagulation (PRP) with the argon laser may slow or stop progression of the disease, but cannot restore vision already lost. Numerous lesions are placed on the periphery of the retina to stop proliferation of the vessels. The characteristics of the argon laser cause absorption in the pigment epithelium and the photoreceptor area. The green wavelength of the argon laser allows deeper penetration of the beam with less damage to surface retinal vessels.

Krypton lasers produce yellow and red light. For work in the macular area this light will spare the macula lutea and be absorbed in the pigmented epithelium. The yellow xanthophyll pigment is contained in the macula and will not absorb yellow light at all and red light very poorly. Red light is used in treatment of subretinal neovascular membranes. It spares surface vessels and the macula, and destroys the pathological subretinal blood supply. Some types of senile macular degeneration (SMD) are very responsive to laser treatment.

Lasers have also been used in different ways to treat glaucoma. In closed angle glaucoma, iridotomy may be performed with the laser to open a channel for fluid flow between the anterior and posterior chambers. Lasers used for this procedure include the argon, Q-switched Nd:Yag, holmium and excimer. There has been a problem with the long-term patency of laser iridotomies, but it is a simple procedure that may be quickly and easily repeated.

Laser trabeculoplasty is used to treat open angle glaucoma. The laser creates multiple lesions around the periphery of the iris into the trabecular meshwork. It thermally shrinks the mesh, creating larger spaces for fluid to flow through. Patients can usually maintain acceptable intraocular pressures with minimal medication. Transscleral approaches to the treatment of open angle glaucoma include the use of flat sapphire probes with low power Nd:Yag laser. The probe is held against the sclera to transmit through it to the angle of the eye.

Pulsed Nd:Yag lasers, of the Q-switched or mode locked variety, produce nonlinear effects at a small focal point. Tremendous peak powers in the millions of watts delivered in an ultrashort burst cause a tiny concussive effect at the 50 μm spot size. The spark seen and crack heard literally snaps apart the membrane. The primary use of photodisruptors is secondary to cataract surgery, for posterior capsulotomy. They are also used for vitreoretinal and glaucoma surgery, and to cut internal sutures.

In posterior capsulotomies, the laser is not actually used to remove the primary cataract. However, on cataracts suspected of being very hard, the surgeon sometimes will crack apart the hard lens in the clinic with the laser before proceeding with conventional surgery. When the diseased lens is removed, the posterior capsule is left intact to provide support for the intraocular lens implant (IOL) and decrease the likelihood of some postoperative complications. Unfortunately, this membrane later becomes clouded in a percentage of patients. Without a laser, an invasive surgical procedure is required to open the clouded membrane. Instead, the laser snaps an opening in the membrane in just a few minutes with several laser shots.

When a photodisrupting laser is used for procedures that may involve a blood supply, such as cutting vitriol strands, an argon laser should be available. If the cold cut of the Nd:Yag creates a small intraocular hemorrhage it is important to be able to cauterize it immediately with the argon. Some ophthalmic Nd:Yag lasers have what is termed a 'free running' or thermal mode. This allows continuous wave operation for photocoagulation. The power is not sufficient to use this mode for other surgical specialties that use continuous wave Nd:Yag lasers.

There is considerable excitement over the use of the excimer laser, in particular the 193 nm argon fluoride, for use in corneal work. The very fine 'non-thermal' cut is consistent and predictable on the micrometer level. In photorefractive keratectomy (PRK) it can reshape the cornea, under computer control, to correct vision. Other cutting and shaping procedures are being investigated, but PRK seems to be proving itself in clinical use. Some hazing is evident during healing, but this seems to be a satisfactory trade-off for those patients undergoing the procedure.

Dacryocystostomy (DCR) is performed using the holmium and Nd:Yag laser and the contact Yag is being used in performing oculoplastic surgery.

Peripheral vascular surgery and coronary laser angioplasty

The use of lasers in this area centers primarily around laser recanalization to open totally or partially occluded blood vessels or tissue welding. Research both preclinical and clinical has been performed with most types of lasers. It is impossible to predict just which type of laser or delivery system will ultimately prove superior.

Tissue fusion of small vessels is occurring experimentally in laboratories and has been used clinically in a few cases. Low power density CO_2 lasers are pulsed on the seam to create an immediate and permanent laser weld with power densities in the broad range of 5–80 W/cm^2. One way to achieve steady, low power density is to use a milliwatt output laser. This produces a very stable output which may be focused to microspot sizes and still retain a sufficiently low power density. This has worked the best for CO_2 lasers. It is also possible to use low power (0.5–3 W) from a conventional CO_2 laser and simply broaden the spot sufficiently to achieve very low power density.

Argon lasers used with handheld fibers at 0.5–3 W output may be used to weld arteriotomies with good results. Argon has also been used to achieve vessel fusion. However, the technique is difficult and not widely practiced. A 1.3 μm Nd:Yag laser, programmed for the correct dosimetry on various tissues, is also being investigated as a better laser alternative for tissue.

Laser recanalization of blood vessels is possible with several laser techniques. Fibers are passed through endoscopes or catheters into the vessel and used to open the blockage. The type of laser used and its delivery system are two separate areas of development. Argon, Nd:Yag, pulsed dye, Ho:Yag and excimer lasers are the front runners for the wavelength. Delivery systems used include bare fibers, hot metal tips, sapphire probes, 'ball tip' fibers, quartz domed fibers including multigang bundles, and handheld needle delivery systems.

The hot metal tips were primarily used with the argon or Nd:Yag lasers as the heat source. These bullet-shaped tips come in a variety of configurations to provide only heating effects at the tip, or a combination of laser/hot tip effect. They may be placed over a guidewire, or used on their own. The laser energy heats the metal tip in the front, so that it vaporizes away obstructions with limited damage to the sidewalls, unless left statically in place against the vessel wall.

The initial overpromotion of this hot tip technique by manufacturers, and poor long-term results have resulted in this technique being widely abandoned. The problems stemmed from the indiscriminate rather than selective use of hot tip fibers. The heat of the tip caused scarring and strictures in vessels and did not appear to help the long-term patency of balloon dilatation. The reaction has been to abandon this technique totally whereas in fact it can be very useful in selected applications. For instance, the treatment of a totally obstructed vessel can be converted from an open procedure into a balloon dilatation by preliminary debulking with the laser. This is best in large vessels such as the iliac to avoid heating of the vessel wall by the hot tip. The Nd:Yag sapphire probes are used in a similar fashion to the hot metal tips, using a combination of focused laser and heat energy.

Excimer lasers have received much attention for laser angioplasty and are now undergoing clinical trials. The 193 nm wavelength of the argon fluoride (ArFl) excimer would be ideal because of the absence of thermal injury. However, fibers are not yet available for this wavelength. The krypton fluoride (KrFl) laser at 248 nm is a possibility and one manufacturer uses this wavelength. Mutagenic effects of 248 nm are still a concern. The xenon chloride (XeCl) laser at 308 nm passes through conventional fibers and is not mutagenic. This XeCl laser will most likely be the excimer of choice for initial vascular work. The light takes just a small 'bite' with each pulse and there is little heating effect on the vessel. The true pulsing characteristics of these lasers contribute as much to the limited thermal effect as the wavelengths themselves. Other pulsed, fiberoptically delivered lasers are also being examined. The pulsed dye laser is one example. The Ho:Yag is the more common laser now being used for these procedures.

The real question here is not the type of laser, but how the delivery system will be incorporated into the catheter. There are both single fiber and multibundle approaches with many variations of the geometry of fiber placement. Some use tiny diverging lenses and some use the bare fibers, with or without thermocouples for temperature control. Studies are in progress in both the peripheral and coronary laser angioplasty fields with these 'cold' lasers and clinical results are awaited. Transluminal myocardial revascularization is currently being evaluated using a very high powered laser (1000 W) during both coronary bypass surgery and the definitive treatment for incapacitating angina on the beating heart. This procedure could potentially be performed through a limited thoracotomy or percutaneously, but clinical results with long-term follow-up are required.

For the treatment of cardiac arrythmias laser fibers delivered percutaneously through a catheter are being used to create destructive lesions in the heart's conduction system for these types of disorders. Lasers provide a very controlled and precise lesion.

Photodynamic therapy is also being used to try and selectively remove atheromatous plaque from diseased vessels. Early experimental work in obtaining a normal intimal surface is encouraging.

Photodynamic therapy

Photodynamic therapy (PDT) involves the use of a photosensitization agent to treat malignant tumors. In this instance a dihematoporphyrin ether (DHE) is the photosensitization drug used and is activated by 632 nm (red) light produced from an argon pumped dye laser. A hematoporphyrin derivative (HpD) was used previously, but the DHE has been found to be more effective. Other photosensitization agents are being investigated. The laser is used because of its ability to produce intense levels of monochromatic light. Other light sources may be used, such as filtered slide projectors, but these are not as effective. The red light from gold vapor lasers is also being examined as an alternative light source.

Fluorescence of some tumors upon illumination with a Wood's lamp was noted as long ago as 1924. This principle was then used as a localization technique by the systemic injection of hematoporphyrin, beginning in 1942. Lipson reported the use of a hematoporphyrin derivative which was shown to have a superior tumor localization to that of hematoporphyrin. The use of HpD then moved from diagnostic to therapeutic applications when Diamond reported in 1972 the destruction of experimental tumors by white light exposed after HpD injection. Dougherty's group at Roswell Park Memorial Institute has been studying the response of a wide variety of malignant tumors in man to PDT with various photosensitizers. He reported complete or partial response in 98% of cutaneous and subcutaneous malignant lesions in 1978.

From the results obtained by various investigators so far, it is clear that PDT with DHE is a valid treatment. The DHE is initially distributed through all the cells but begins to clear out of normal tissue after several hours. An initial dose is injected as a single intravenous bolus. A maximum difference in concentration levels between tumor cells and normal cells occurs in about three days. Normal tissue retains some DHE and this is complicated by the fact that different tissues retain different concentrations. Skin, liver, kidney and spleen hold onto the DHE longer than other tissues. Bronchial mucosa retains one of the lowest concentrations so endobronchial tumors are among the easiest to treat.

The goal of dosimetry is to calculate dosages of drug and light so that activation of the higher concentrations of DHE occurs in cancer cells while remaining below the necessary threshold to activate DHE in normal tissue. This is controlled by dosage and timing of the DHE injection, color, intensity and distribution of the light, and its method of delivery. The technique is illustrated diagrammatically in Fig. 5.1, and a patient undergoing treatment is shown in Plate I.

Until recently, most patients treated with this form of therapy have previously exhausted the full gamut of conventional therapies of surgery, radiation, immunotherapy and chemotherapy. Results have been encouraging enough for a few investigators to utilize PDT for some early lesions as the primary form of treatment.

The absorption spectrum of DHE has peaks which may be

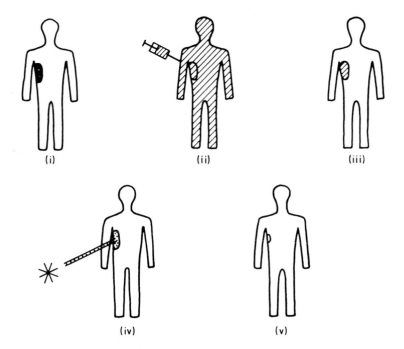

Fig. 5.1 Diagrammatic and simplified illustration of photoradiation therapy: (i) the patient with tumor in chest; (ii) hematoporphyrin derivative (HpD) injected intravenously and taken up by the whole body; (iii) HpD selectively retained by the tumor after about three days; (iv) HpD photoactivated by 630 nm laser light; (v) toxic products produced in (iv) destroy tumor, leaving normal tissues undamaged.

utilized to activate the drug. Blue light of about 405 nm is absorbed most strongly but a lesser absorption peak also exists in the range of red light at 630 nm. The red light, however, will penetrate tissue much better than the blue. Red light will scatter up to approximately 2 cm through skin. The amount of light energy delivered to an area is measured with a radiometer and time is calculated to deliver doses of between 25 J and 150 J/cm^2, depending on the tumor.

The drug is activated by the light probably through singlet oxygen production. The vessels of the tumor's blood supply are implicated strongly in the phototoxic process. Gross tissue effects proceed from moderate or severe edema to complete necrosis of the tumor exhibited by a black eschar.

PDT is still investigational and has not been proven curative for most malignancies. It will in all probability be used as an adjuvant therapy and, for some neoplasms, the primary form of treatment. Its mechanism is independent of previous or subsequent chemotherapy and/or X-ray therapy.

There are side effects and complications to this form of treatment. Patients may undergo an increased photosensitivity of the skin to sunlight for about one month after treatment. Full thickness necrosis of tumors of the intestinal tract may lead to fistula formation. Endobronchial treatments may cause production of gelatinous secretions and edema of the airway causing obstructions. Post-treatment hemorrhage is possible following necrosis of tumor and any involved vessels.

Current technology with lasers and fiberoptics makes it feasible to deliver high intensity light to almost any site in the body, at surgery, or through endoscopes and percutaneously via needles. PDT may be used to treat malignancies that are not responsive to current modalities. It may be used to treat nonresectable lesions of the pancreas or brain. Current applications also include pulmonary, urologic, gastrointestinal, gynecologic, ENT, and dermatologic tumors. Concurrent therapies with systemic treatment, ionizing radiation or hyperthermia are possible since PDT is a local treatment.

One of the exciting recent developments is the use of PDT to treat clinical and subclinical HPV infections seen as extensive condylomata in gynecology practices. CO_2 lasers are now used advantageously in treatment of resistant infections, but the use of

PDT would be better for patients and physicians. The DHE localizes in the HPV virus, allowing selective destruction of all the latent virus in a painless treatment. Postoperatively the patient may be inflamed and sore, but not hurt like after an extensive CO_2 vaporization.

PDT may also be used in the processing of blood products to clear HIV, hepatitis, and other viruses from the products. Viruses seem to take up the photosensitizers to a high degree whereas blood constituents do not. The blood may be passed through a chamber which illuminates the blood with high intensity light of specific wavelengths (blue or red). Viruses are thus cleared from the blood samples.

Laserthermia or interstitial local hyperthermia

The Nd:Yag laser has been used in combination with contact interstitial endoprobes and temperature sensors to treat cancer both experimentally and clinically. The tumor tissue is heated between 42° and 44°C for 20–30 minutes using computer controlled equipment. Over several days, tumors necrose leaving normal tissue behind.

Applications include open, endoscopic and percutaneous placement of laser probes. Combination therapy with PDT appears synergistic. Lesions being treated on an investigational basis in the USA include tumors of the esophagus, stomach, brain, liver, pancreas, breast and oral cavity.

Biostimulation

Though not one of the main uses of lasers in medicine, biostimulation does deserve mention. Very low levels of laser light – typically red or very near infrared – can induce physiologic, nondestructive changes in cells. These can stimulate or inhibit certain cells and even alter internal enzyme activity within a cell. Veterinarians have been using biostimulation for a number of years to treat a variety of conditions in animals, most notably tendinitis in race horses.

These are sometimes referred to as 'soft' or 'cold' lasers. They are being investigated for the treatment of chronic headaches, back pain, joint and other pain. Wound healing is also being investigated. They have been shown to alter nerve condition rates and accelerate hair growth from an existing follicle. Acupuncturists use lasers instead of needles to avoid problems of needle sterilization.

The current problem with this field is that it is still in the developmental stage and claims are sometimes made by individuals of questionable credibility. However, it is important not to dismiss the procedure, but to await documented scientific randomized prospective studies before making a decision on the validity of biostimulation. Soft lasers do produce some very subtle physiologic effects but the question remains unanswered as to how and when to apply this information clinically.

6   Laser safety

General points

Hazards associated with the use of surgical laser systems require appropriate safety precautions and policies. It is beyond the scope of this section to delineate fully the potential biological and ocular hazards associated with the full spectrum of laser wavelengths. A suggested reading list is included at the end of the book for those wishing more comprehensive information.

We will discuss practical safety considerations as they relate to the surgical or outpatient environment with CO_2 argon (or KTP) and Nd:Yag laser systems. There is a great deal of overlap in safety practices with these three lasers, with eye protection being the biggest differentiating factor.

Laser manufacturers are required under the ANSI-Z–136.1 standard in the USA and the BSI standard BS 4803 in the United Kingdom to provide certain design and engineering safety measures in their equipment. In the USA, expressed or implied warranties of merchantability may not be abrogated by a manufacturer if service is provided on the equipment by third party service agents. In other countries equivalent conditions must be observed. Safety requirements include provision for such items as key interlocks and visual or audible warning systems. All of the required measures are in place on commercially available lasers and the user need not be concerned with complying with these requirements.

While there is broad international agreement over safety levels and practices involving the use of laser, differences exist between national administrative mechanisms for accrediting doctors for laser work.

In the United Kingdom, a local body such as the Radiological Safety Committee is concerned with laser safety policy, but does not take upon itself responsibility for accrediting laser users. However, the British Medical Laser Association (BMLA) has been asked to consider accreditation in individual cases.

In the USA, no medical or governmental agency credentials physicians for laser use. Doctors are credentialed through their medical license and respective specialty boards to practice medicine. Laser is simply a tool used in their practice of medicine. The American Society for Laser Medicine and Surgery (ASLMS) of Wausau Wisconsin also does not credential physicians, however they do have available recommendations on training suggested before clinical laser use.

The American Board of Laser Surgery offers board examinations on laser use and provides credentialing certificates. This is a private company and is not associated with the American Medical Association (AMA) and its specialty credentialing boards. The objective of this company is to provide some type of evaluation process to hospitals as a basis for granting laser privileges to physicians. In the sense that it has identified a deficit area in evaluation of physicians' knowledge of laser use, its efforts are worthwhile. Board certification by this company is not required for laser use in the United States. The question of physician evaluation will best be confronted when the true medical specialty boards start including laser-related questions in their board examinations.

Common practice includes the establishment of a Laser Safety Committee with representatives from each of the interested specialties, anesthesiology, operating room director, and administration. This committee can adopt their own credentialing standards for surgeons and establish formal laser safety policies and procedures. Initially this committee is formed to evaluate laser equipment for purchase.

Training is the single most important factor in the safe use of any laser. This applies to surgeons, laser specialists, and operating room personnel.

Customary credentialing standards for physicians include a 12–16 hour hands-on training experience with the laser. This is broken down into three primary areas of experience.

1. Fundamentals of laser surgery (theory)
2. Laboratory hands-on animal or cadaver work (in most cases)
3. Clinical orientation to procedures in that specialty

Additionally, some committees require an in-house preceptor for the surgeon's first one or two cases. Once a surgeon is credentialed on a particular type of laser, he or she should not necessarily have to repeat the entire course for different lasers. Some type of supervised, hands-on orientation to the new type of laser should be sufficient.

Several nurses or surgical technicians may be selected to receive thorough training in the operation and safety practices of the laser. These will be the laser specialists. A laser manufacturer's inservicing of how to operate that particular unit should be complemented with more thorough instruction in the theory of the primary surgical lasers, nursing considerations and safety practices.

Lasers operated in the simpler environment of an office or clinic may not require dedicated specialists to run the laser. Often the physician or staff in the room can easily do this.

In surgery, laser specialists operate the equipment. The circulator should not be used for this purpose if they must frequently leave the room. The function of the laser specialist is to ensure that hospital personnel follow recommended safety practices, wear the correct protective eyewear, and follow established procedures. The laser should always be placed in the standby position when the surgeon is not using it. If applicable, the laser nurse will also ensure that the physician is on the approved list for laser privileges from the safety committee. The laser specialist will precheck the laser to assure its proper operation, and may maintain a laser log for all cases. This log does not become part of the patient's chart. The physician will chart whatever laser remarks are appropriate, if any, in their operative report. Documentation of periodic maintenance, power calibration and any service work performed is very important.

Operating room personnel need to be instructed in safety procedures relevant to the type of laser used, but need not be taught detailed operation of the units. Warning signs should be placed on all entries to the surgical suite before the laser is operated.

The signs, which are commercially available, should state the type and class of laser used, and its maximum power output.

The keys to the laser should be controlled by the laser specialist, or unit management. The keys should not be left in the lasers. This will control unauthorized access.

Electrical hazards are shared by all types of lasers, the same as for any electrical equipment. Lasers contain high voltage power supplies and panels should only be removed by trained personnel. Liquids should not be placed on the units because of the potential of spills and resulting short circuiting of the laser.

The radiation hazards of the laser, posted on the warning signs, have nothing to do with X-rays or any type of diffuse ionizing radiation around the equipment. The light itself is the radiation referenced. Apart from eye safety, this beam is only a hazard on direct impact, which would cause a thermal burn. Women in any stage of pregnancy may work around conventional laser units with no adverse effects. An X-ray laser does exist, but is a star wars variety high power laser of interest primarily to the military.

CO_2 lasers

This is the most common laser used in an operating room. This wavelength is outside the retinal hazard region because diffuse reflections are not transmitted into the eye. It does present the potential for corneal or scleral burns (or anywhere on the body) when the beam is sufficiently focused. The effect is both immediate and painful. For this reason safety glasses are required. Either glass or plastic will absorb a diffuse CO_2 laser beam. Common practice in a surgery setting is to wear one's own corrective glasses. Side shields are available for additional protection if desired, but are not really necessary. Personnel working in the field, close to the focused beam may simply wish to wear the plastic safety glasses over their own glasses if the additional side protection is desired. Some outpatient clinics have opted to make the glasses available, but optional, for personnel in the room because of the relative safety of using these units, especially when attached to a colposcope. Contact lenses of any type are inadequate. In the United Kingdom, ordinary safety glasses would be considered inadequate, as they do not conform to the requirements of the Protection of Eye Regulations 1974. The lenses in any microscope or rigid endoscope serve as the protection for the operating surgeon.

The patient's eyes must also be protected if they are in a position that exposes them to the laser. Wet sponges taped in place over the eyes if the patient is asleep, or safety glasses if they are awake, will provide protection.

Sponges and drapes are also kept moist for fire protection. These materials are soaked only in the immediate surgical field. Dry sponges will flame immediately upon impact with the CO$_2$ laser. A container of water or saline should always be kept available to douse a flame if needed. The normal pan of irrigation solution will serve this purpose.

Most paper drapes available in the USA are fire resistant and may be safely used with the laser. Cloth drapes may have moist sponges draped at the perimeter of the field. A drape will flame only if it is directly impacted with the laser beam.

Do not use the laser in the presence of flammable preparative solutions or drying agents. This is most important for the CO$_2$ laser but would apply to any type. These preparative solutions may be used on the patient, but the area should be dry before lasing.

During oral, nasopharyngeal or laryngotracheal surgery, special precautions must be taken to avoid an airway explosion and blowtorch type of fire. A polyvinylchloride (PVC) endotracheal tube must *never* be used. Alternatives include Norton flexible metal tubes, red rubber tubes wrapped with foil tape, or laser-resistant tubes available from several manufacturers.

When the Norton tubes are used, no distal cuff is placed, so that no flammable materials rest in the airway. This provides complete protection against fire but is an open ventilation system. If sized well, they provide a good fit and leak very little.

Red rubber tubes are wrapped in an overlapping spiral fashion from the cuff up with reflective foil tape. Each batch of tape should be tested to assure that the laser will not penetrate. After intubation, the inflated cuff is packed off well with moist cottonoids. The cuff may be inflated with saline and methylene blue dye. If the cuff were inadvertently ruptured, the blue saline would soak through and the surgeon would be aware of the deflation.

Manufacturer's recommendations should be followed in using

laser-resistant type tubes. It would still be helpful to inflate the cuff with fluid and pack off the end.

None of the tubes should be tightly taped in place during these procedures so that they may be pulled quickly if a fire starts. The tube may be marked with a distance marker to maintain the correct distance.

The laser must obviously never be used in the presence of explosive anesthetics.

Methane gas is also flammable. Whenever lasing is done up into the rectum, it should be packed or covered with a moist towel or sponge. Failure to do so could cause serious internal injury to the patient and/or facial burns to the surgeon.

During microlaryngoscopies, or when otherwise exposing the patient's face to the laser beam, the entire face should be draped with wet towels. The only exposed area should be the oral cavity. Care should be taken to protect the teeth with bite blocks or soaked sponges. A CO_2 laser impact on tooth enamel will create a small, permanent, black hole in the tooth.

Special instrumentation is available that has been blackened or anodized to reduce the danger of the beam reflecting off the instrument and striking other objects. These instruments are an important safety precaution for microlaryngoscopy. They are of benefit, but not critical, in others types of CO_2 laser surgery. It is the diffuse matt finish of the instrument that reduces reflection rather than the black color. When regular retractors are being used, they may be covered with moistened sponges for protection. Teflon-coated instruments should not be used in a laser field. They produce toxic fumes when lased.

Pyrex, quartz, or more often titanium rods are routinely used in gynecologic laser surgery as fine dissecting rods and backstops. Glass rods should not be used because they will shatter, leaving tiny glass fragments in the field.

Nd:Yag lasers

Safety precautions center around eye safety and care of endoscopes. Fire precautions must still be taken, but this is not as much of a hazard here as with the CO_2 laser.

It is critically important that all personnel be wearing the appropriate safety glasses *before* they enter the room when the laser is being fired. Regular glasses will not suffice. Most of the Nd:Yag safety glasses are green or green–gray colored, though some of the newest are almost totally clear. All safety glasses are marked as to the wavelength covered, as well as the optical density (darkness) of the material. It is very important to look at the wavelength markings on these safety glasses before putting them on. Some of the newer Nd:Yag safety glasses are clear in appearance. If one did not look at the markings for 1060 nm (or 1.06 μm) to cover the Nd:Yag, it would be possible to get the glasses mixed up. Safety goggles are designed to fit over ones own corrective glasses. Prescription laser safety glasses are available for purchase. Eyepiece filters are available to fit over the end of endoscopes so that the endoscopist need not wear safety glasses. It is critically important that anyone viewing through a teaching head also has eye protection. It is a common safety practice for all personnel in the room to wear protective eyewear during endoscopy, even though they are not actually at risk when the laser is fired only in this closed environment. With the use of contact probes or sculpted fibers, the combination of the probe and low powers significantly reduces potential eye hazards, however protective eyewear is still recommended. The lasers are reasonably safe, even with no protective eyewear, as long as you do not look directly into the fiber as it is fired. The problem is the inability to predict when a problem could occur, which is why protective eyewear is important.

As with any laser, it is possible to install door interlocks that will automatically shut off the laser when the door is opened. The authors discourage the use of these door interlocks in a surgery setting and instead reassert the fact that training is the single most important factor in the safe use of any laser. Windows in the room must be covered so that observers are protected from unintentional viewing. Panels may be made for the windows, or towels may simply be taped over them. Patient's eyes are protected by covering with towels or sponges. Awake patients should wear protective eyewear.

Operating microscopes do not provide eye protection for the Nd:Yag laser and the surgeon must either wear protective eyewear, or a filter incorporated into the manipulator device. Room personnel must all wear protective eyewear.

Fibers which allow coaxial delivery of cooling gas can be deadly to the patient if used inappropriately. The lasers most frequently use compressed air as the cooling gas, though it is possible to use other gases or even fluid. Problems have occurred most recently when these fibers are used in the uterus for hysteroscopy. The fibers have been placed onto tissue to cut or ablate, thus opening the vasculature of the uterus and injecting large volumes of high velocity compressed air. Massive embolisms can result. These fibers are not dangerous in themselves. It is the inappropriate placement of these gas fibers directly into tissue which creates the problems. Though people seemed to be obsessed with the potential eye hazards of this laser, the gas fiber hazard is a very real danger compared with the unlikelihood of inadvertent eye damage.

It is possible to cause extensive damage, or total destruction, to flexible endoscopes by firing the laser with the fiber not fully extended from the channel. This is possible because one can see the guide light on the target while the fiber is still within the scope. As a matter of policy, it is important for the surgeon to have the tip of the fiber in his field of view before firing the laser. A distance of 1.0–2.0 cm is an adequate distance from the scope.

Because of the high degree of back scatter with the Nd:Yag laser, and the dark color of the tip of most endoscopes, it is possible to melt this tip and even cause a small fire if high concentrations of oxygen are used, as in bronchoscopy. The further away from the tip the fiber is placed the better.

A fire hazard also exists when lasing an airway tumor while ventilating with high concentrations of oxygen. Fat in a tumor may flare from the laser and, in the presence of oxygen, cause a small flash. Though undesirable, this does not pose the risk and types of problems associated with an endotracheal tube fire from the CO_2 laser. Entrainment of air into the ventilating gas is sometimes used to maintain lower (less than 40%) oxygen concentrations. The use of pulse oxymeters allows the lowest, safest oxygen concentrations.

In ophthalmic applications the beams of the Nd:Yag laser (photodisruptors) are generally focused at a large angle of convergence, so that beyond the focus the beam intensity drops quickly. However, reflections from the surface of the cornea or a contact lens can still be hazardous up to a meter or so away, depending on the curvature of the reflecting surface. Therefore, it is impor-

tant that attendant staff wear safety glasses when treatments are
in progress.

Holmium:Yag lasers

Burn and fire hazards are similar with the Ho:Yag as with other
fiber lasers, that is, they are reasonably safe unless buried directly
into the object.

Eye hazards are somewhat different with this wavelength of
around 2.1 μm because it is partially transmitted into the eye. It
does not pose a direct corneal hazard, but does pose a direct
retinal hazard. In this sense it is one of the safer lasers in terms
of eye safety. It is absorbed in the vitreous of the eye and its long-
term effects here are unknown. Discretion advises the use of
safety glasses when the beam is exposed to viewing. Its use internally in arthroscopy or cardiovascular applications would mean that
it is technically eyesafe as long as the laser is activated only when
the fiber is placed internally. The same is also true of other lasers.

It is important to keep the fiber tip away from the viewing lens
of the arthroscope when performing those procedures. Since joints
are such small spaces to begin with, it is easy to put the fiber right
up against the scope. This will damage or destroy the scope.

Pulsed dye lasers

These lasers produce very bright bursts of visible light up to
40 000 W of power. The dermatologic units (yellow light) are used
externally by definition. It is *very* important to wear the correct
eyewear when using these lasers. Apart from eye safety, the lasers
are very safe to use.

Argon, KTP and other 532 nm lasers

Essentially, the wavelengths on these lasers are close enough that
safety glasses and precautions are identical. The exception is a
special safety glass that the KTP manufacturer has designed to
such a narrow bandwidth to be useful for only this laser. The
discussion of argon also implies application to KTP.

All laser procedures are safe for both patient and personnel
provided proper training is acquired, necessary precautions are

observed and equipment is maintained in a proper operating order. Commonsense is the most important factor in safety.

   . . . and there was light! ! !

Further reading

Absten, G.T. (1987) *Evaluation of Surgical Laser Technology.* Advanced Laser Services Corp., Columbus, OH.

Absten, G.T. (1988) *The Myth of Lasers in Medicine.* Advanced Laser Services Corp., Columbus, OH.

Absten, G.T. (1991) *Fundamentals of Electrosurgery.* Advanced Laser Services Corp., Grove City, OH.

American National Standard for the Safe Use of Lasers. ANS12136,1 (1990) American National Standards Institute, New York.

Apfelberg, D. (1987) *Evaluation and Installation of Surgical Laser Systems.* Springer-Verlag, New York.

Baggish, M.S. (1990) *Clinical Practice of Gynecology – Endoscopic Laser Surgery.* Elsevier, New York.

Calder, N. (1979) *Einstein's Universe.* Viking Press, New York.

Dixon, J.A. (1987) *Surgical Application of Lasers*, 2nd edn. Year Book Medical Publishers, Chicago, IL.

Dorsey, J.H. (ed.) (1991) *Obstetrics and Gynecology Clinics of North America.* W.B. Saunders Co., Harcourt Brace Jovanovich Inc., Philadelphia, PA.

Einstein, A. (1954) *Ideas and Opinions*, Bonanza Books, New York.

Einstein, A. (1961) *Relativity.* Bonanza Books, New York.

Fuller, T.A. (1987) *Surgical Lasers, A Clinical Guide.* Macmillan Publishing Co., New York.

Goldman, L. (1991) *Laser Non-Surgical Medicine.* Technomic, PA.

Hallmark, C. (1979) *Lasers, The Light Fantastic.* Tab Books, Blue Ridge Summit, PA.

Hecht, J. (1988) *Understanding Lasers*, Howard W. Sams and Co., A Division of Macmillan Inc., New York.

JGM Associates Inc. (1991) *Endoscopic Therapy Applications of Advanced Solid State Lasers*. JGM Associates Inc., Burlington, MA.

Joffe, S.N. (1988) *Lasers in General Surgery*. Williams and Wilkins, Baltimore, MD.

Joffe, S.N. and Oguro, Y. (1987) *Advances in Nd:Yag Laser Surgery*. Springer-Verlag, New York.

Lanzafame, R. and Hinshaw, J.R. (1988) *Atlas of CO₂ Laser Surgical Techniques*. Ishiyaku EuroAmerica, St Louis, MO.

Laser Focus/Penwell Publications (1991) *Medical Laser Buyers Guide*. Littleton, MA.

Mackety, C.J. (1989) *Perioperative Laser Nursing – A Practical Guide*, 2nd edn. Laser Centres of America, Ohio.

Martin, D.C., Absten, G.T., Levinson, C.J. and Photopulos, G.J. (1986) *Intra-abdominal Laser Surgery*. Resurge Press, Memphis, TN.

McLaughlin, D.S. (1991) *Lasers in Gynecology*. J.B. Lippincott, Philadelphia, PA.

Masten, L.B. and Masten, B.R. (1981) *Understanding Optronics*. Texas Instruments, Dallas, TX.

O'Connor, D.T. (1987) *Current Reviews in Obstetrics and Gynecology, Endometriosis*. Churchill Livingstone, Longman Group, London, UK.

Ratz, J.L. (1986) *Lasers in Cutaneous Medicine and Surgery*. Year Book Medical Publishers, Chicago, IL.

Sanfilippo, J. S. and Levine, R.L. (1988) *Gynecologic Endoscopy, Pelviscopic Surgery and Laser Laparoscopy*. Springer-Verlag, New York.

US Dept of Health, Education and Welfare Public Health Service Food and Drug Administration, *Biological Bases for and Other Aspects of a Performance Standard for Laser Products*. HEW Publications (FDA) 80–8092.

Index

Index **89**